Air Conditioning and Heat Pumps

Home Inspection

This publication is designed to provide accurate and authoritative information in regard to the subject matter covered. It is sold with the understanding that the publisher is not engaged in rendering legal, accounting, or other professional service. If legal advice or other expert assistance is required, the services of a competent professional person should be sought.

President: Roy Lipner
Publisher and Director of Distance Learning: Evan M. Butterfield
Senior Development Editor: Laurie McGuire
Content Consultant: Alan Carson
Acting Editorial Production Manager: Daniel Frey
Creative Director: Lucy Jenkins
Graphic Design: Neglia Design Inc.

© 2003
Carson Dunlop & Associates Limited. All rights reserved.

Published by Dearborn™ Real Estate Education
a Division of Dearborn Financial Publishing, Inc.®
a Kaplan Professional Company®
30 South Wacker Drive
Chicago, IL 60606-7481
www.dearbornRE.com

All rights reserved. The text of this publication, or any part thereof, may not be reproduced in any manner whatsoever without written permission in writing from the publisher.

Printed in the United States of America.

03 04 05 10 9 8 7 6 5 4 3 2 1

INTRODUCTION

Welcome! This home inspection training program has two primary goals:

- To provide you with a sound introduction to the components, materials and mechanics of house systems that you will encounter and evaluate as a home inspector;

- To provide you with a solid understanding of inspection processes, strategies and standards of practice that will help define the scope of your inspections.

We hope you enjoy this training program and develop a good understanding of the various house systems as you proceed. Good luck!

FEATURES OF THIS PROGRAM

This program is structured to help you learn and retain the key concepts of home inspection. It also will help you form a set of best practices for conducting inspections. A number of features are included to help you master the information and put your knowledge into practice:

- Topics are organized into evenly paced Study Sessions. Each Session begins with learning objectives and key words to set up the important concepts you should master by the end of the Session. Each Session concludes with Quick Quizzes to help you test your understanding. Answers to Quick Quizzes are provided so you can check your results.

- Scope and Introduction sections present the ASHI® (American Society of Home Inspectors) Standards of Practice for each major topic. Standards help you define a professional, consistent depth and breadth for your inspections.

- An Inspection Checklist at the end of each section summarizes the important components you will be inspecting and their typical problems. You can use this as a set of field notes during your own inspections.

- The Inspection Tools list will help you build your toolkit of "must have" and optional tools for the job.

- An Inspection Procedures section provides some general guidelines to conducting your inspection of each major house system. This feature will help you develop a methodology to complement your technical knowledge.

- Field Exercises give you an opportunity to turn your knowledge into real world experience.

SUMMARY

The road we have paved for you is designed to be easy and enjoyable to follow. We trust it will lead you quickly to your destination of success in the home inspection profession.

ACKNOWLEDGMENTS

Thanks to Kevin O'Malley for his inspiration, advice and guidance. Thanks also to James Dobney for his invaluable input and encouragement. Special thanks are extended to Dan Friedman for his numerous and significant contributions.

We are grateful for the contributions of: Duncan Hannay, Richard Weldon, Peter Yeates, Tony Wong, Graham Clarke, Ian Cunliffe, Joe Seymour, Charles Gravely, Graham Lobban, Dave Frost, Gerard Gransaull, Jim Stroud, Diana DeSantis, David Ballantyne, Shawn Carr and Steve Liew.

Special thanks are also extended to Susan Bonham, Dearbhla Lynch, Lucia Cardoso-Tavares, Jill Brownlee, Ida Cristello and Rita Minicucci-Colavecchia who have brought everything together. Thanks also to Jim Lingerfelt for his invaluable editing assistance.

SECTION ONE

▶ AIR CONDITIONING

	1.0	OBJECTIVES 5
STUDY SESSION 1	2.0	SCOPE AND INTRODUCTION 8
	2.1	Scope 8
	2.2	Introduction 16
	3.0	THE BASICS 18
STUDY SESSION 2	4.0	AIR CONDITIONING CAPACITY 42
	4.1	Introduction 42
	4.2	Conditions 45
	4.2.1	Undersized 45
	4.2.2	Oversized 48
	5.0	COMPRESSOR 49
	5.1	Introduction 49
	5.2	Conditions 51
	5.2.1	Excess noise/vibration 51
	5.2.2	Short cycling or running continuously 53
	5.2.3	Out of level 54
	5.2.4	Excess electric current draw 55
	5.2.5	Wrong fuse or breaker size 56
	5.2.6	Electric wires too small 56
	5.2.7	Missing electrical shutoff 56
	5.2.8	Inoperative 57
	5.2.9	Inadequate cooling 58
STUDY SESSION 3	6.0	CONDENSER COIL (OUTDOOR COIL) 64
	6.1	Introduction 64
	6.2	Conditions 64
	6.2.1	Dirty 65
	6.2.2	Damaged/leaking 65
	6.2.3	Corrosion 66
	6.2.4	Clothes dryer or water heater exhaust too close to condenser 66
	7.0	WATER-COOLED CONDENSER COIL 67
	7.1	Introduction 67
	7.2	Conditions 68
	7.2.1	Leakage 69
	7.2.2	Coil cooled by pool water 70
	7.2.3	No backflow preventer (anti-siphon device) 70
	7.2.4	Low plumbing water pressure 71

SECTION ONE: AIRCONDITIONING

	8.0	**EVAPORATOR COIL (INDOOR COIL)**	**72**
	8.1	Introduction	72
	8.2	Conditions	75
	8.2.1	No access to coil	75
	8.2.2	Dirty	76
	8.2.3	Frost	77
	8.2.4	Top of evaporator dry	78
	8.2.5	Corrosion	78
	8.2.6	Damage	78
	8.3	Expansion device (metering device)	79
	8.3.1	Conditions	80
	9.0	**CONDENSATE SYSTEM**	**82**
	9.1	Condensate drain pan (tray)	82
	9.1.1	Introduction	82
	9.1.2	Conditions	82
	9.2	Auxiliary condensate drain pan	83
	9.3	Condensate drain line	84
	9.3.1	Introduction	84
	9.3.2	Conditions	85
	9.4	Condensate pump	88
	9.4.1	Introduction	88
	9.4.2	Conditions	89
	10.0	**REFRIGERANT LINES**	**90**
	10.1	Introduction	90
	10.2	Conditions	93
	10.2.1	Leaking	93
	10.2.2	Damage	95
	10.2.3	Missing insulation	95
	10.2.4	Lines too warm or too cold	96
	10.2.5	Lines touching each other	96
STUDY SESSION 4	**11.0**	**CONDENSER FAN (OUTDOOR FAN)**	**102**
	11.1	Introduction	102
	11.2	Conditions	102
	11.2.1	Excess noise/vibration	103
	11.2.2	Inoperative	103
	11.2.3	Corrosion or mechanical damage	103
	11.2.4	Obstructed air flow	104

SECTION ONE: AIR CONDITIONING

12.0	**EVAPORATOR FAN (INDOOR FAN, PLENUM FAN, BLOWER OR AIR HANDLER)**	**105**
12.1	Introduction	105
12.2	Conditions	107
12.2.1	Undersized blower or motor	107
12.2.2	Misadjustment of belt or pulley	108
12.2.3	Excess noise/vibration	109
12.2.4	Dirty fan	110
12.2.5	Dirty or missing filters	110
12.2.6	Inoperative	111
12.2.7	Corrosion/mechanical damage	111
13.0	**DUCT SYSTEM**	**112**
13.1	Supply and return ducts	112
13.1.1	Conditions	112
13.2	Duct insulation	124
13.2.1	Conditions	125

STUDY SESSION 5

14.0	**THERMOSTATS**	**133**
14.1	Introduction	133
14.2	Conditions	134
14.2.1	Inoperative	134
14.2.2	Poor location	135
14.2.3	Not level	136
14.2.4	Loose	137
14.2.5	Dirty	137
14.2.6	Damaged	137
14.2.7	Poor adjustment/calibration	138
15.0	**LIFE EXPECTANCY –**	**138**
16.0	**EVAPORATIVE COOLERS**	**140**
16.1	Introduction	140
16.2	Conditions	143
16.2.1	Leaking	143
16.2.2	Pump or fan inoperative	143
16.2.3	Rust, mold and mildew	144
16.2.4	No air gap on water supply	144
16.2.5	No water	144
16.2.6	Poor support for pump and water system	145
16.2.7	Louvers obstructed	145
16.2.8	Missing or dirty air filter	145
16.2.9	Cabinet too close to grade	145
16.2.10	Cabinet or ducts not weather-tight	146
16.2.11	Electrical problems	146
16.2.12	Duct problems	146
16.2.13	Excess noise or vibration	147
16.2.14	Clogged pads	147

SECTION ONE: AIR CONDITIONING

17.0		**WHOLE-HOUSE FAN**	**147**
	17.1	Introduction	147
	17.2	Conditions	149
	17.2.1	Inoperative	149
	17.2.2	Inadequate attic venting	150
18.0		**INSPECTION TOOLS**	**154**
19.0		**INSPECTION CHECKLIST**	**155**
20.0		**INSPECTION PROCEDURES**	**157**

FIELD EXERCISE 1

21.0	**BIBLIOGRAPHY**	**167**
	ANSWERS TO QUICK QUIZZES	**168**

SECTION TWO

► HEAT PUMPS

	1.0	OBJECTIVES 2
STUDY SESSION 1	2.0	INTRODUCTION 5
	3.0	HEAT PUMPS IN THEORY 5
	3.1	The Concept..5
	3.2	The Mechanics 6
	3.3	Co-Efficient of performance (COP)..10
	3.4	When the heat pump cannot keep up 12
	3.5	Humidity is the enemy 14
	3.6	Cost effectiveness 15
	3.7	One little twist..15
STUDY SESSION 2	4.0	HEAT PUMPS IN PRACTICE 22
	4.1	Undersized or oversized..22
	4.2	Heat pumps are similar to air conditioners 23
	4.3	Differences between air conditioners and heat pumps 23
	4.4	Heat pumps differ from gas and oil furnaces..............................26
	4.5	Defrost cycle 26
	4.6	Identifying heat pumps 28
	4.7	Conditions 29
	4.7.1	Oversized for cooling and undersized for heating 33
	4.7.2	Inoperative in heating or cooling mode 33
	4.7.3	Poor outdoor coil location...34
	4.7.4	Coil iced up 35
	4.7.5	Airflow problems in house 36
	4.7.6	Back-up heat problems 37
	4.7.7	Old 39
	4.8	Other types of heat pumps 39
	4.8.1	Water-to-air system..39
	4.8.2	Earth-to-air system...40
	4.8.3	Solar systems 42
	4.8.4	Bivalent systems 43
	5.0	INSPECTION TOOLS 51
	6.0	INSPECTION CHECKLIST..52
	7.0	INSPECTION PROCEDURE 54
FIELD EXERCISE 1		
	8.0	BIBLIOGRAPHY...64
		ANSWERS TO QUICK QUIZZES..65

ix

1 AIR CONDITIONING

Air Conditioning & Heat Pumps
MODULE

SECTION ONE: AIR CONDITIONING

▶ TABLE OF CONTENTS

	1.0	OBJECTIVES	5
STUDY SESSION 1	2.0	SCOPE AND INTRODUCTION	8
	2.1	Scope	8
	2.2	Introduction	16
	3.0	THE BASICS	18
STUDY SESSION 2	4.0	AIR CONDITIONING CAPACITY	42
	4.1	Introduction	42
	4.2	Conditions	45
	4.2.1	Undersized	45
	4.2.2	Oversized	48
	5.0	COMPRESSOR	49
	5.1	Introduction	49
	5.2	Conditions	51
	5.2.1	Excess noise/vibration	51
	5.2.2	Short cycling or running continuously	53
	5.2.3	Out of level	54
	5.2.4	Excess electric current draw	55
	5.2.5	Wrong fuse or breaker size	56
	5.2.6	Electric wires too small	56
	5.2.7	Missing electrical shutoff	56
	5.2.8	Inoperative	57
	5.2.9	Inadequate cooling	58
STUDY SESSION 3	6.0	CONDENSER COIL (OUTDOOR COIL)	64
	6.1	Introduction	64
	6.2	Conditions	64
	6.2.1	Dirty	65
	6.2.2	Damaged/leaking	65
	6.2.3	Corrosion	66
	6.2.4	Clothes dryer or water heater exhaust too close to condenser	66
	7.0	WATER-COOLED CONDENSER COIL	67
	7.1	Introduction	67
	7.2	Conditions	68
	7.2.1	Leakage	69
	7.2.2	Coil cooled by pool water	70
	7.2.3	No backflow preventer (anti-siphon device)	70
	7.2.4	Low plumbing water pressure	71

SECTION ONE: AIR CONDITIONING

8.0	**EVAPORATOR COIL (INDOOR COIL)**	72
8.1	Introduction	72
8.2	Conditions	75
8.2.1	No access to coil	75
8.2.2	Dirty	76
8.2.3	Frost	77
8.2.4	Top of evaporator dry	78
8.2.5	Corrosion	78
8.2.6	Damage	78
8.3	Expansion device (metering device)	79
8.3.1	Conditions	80
9.0	**CONDENSATE SYSTEM**	82
9.1	Condensate drain pan (tray)	82
9.1.1	Introduction	82
9.1.2	Conditions	82
9.2	Auxiliary condensate drain pan	83
9.3	Condensate drain line	84
9.3.1	Introduction	84
9.3.2	Conditions	85
9.4	Condensate pump	88
9.4.1	Introduction	88
9.4.2	Conditions	89
10.0	**REFRIGERANT LINES**	90
10.1	Introduction	90
10.2	Conditions	93
10.2.1	Leaking	93
10.2.2	Damage	95
10.2.3	Missing insulation	95
10.2.4	Lines too warm or too cold	96
10.2.5	Lines touching each other	96

STUDY SESSION 4

11.0	**CONDENSER FAN (OUTDOOR FAN)**	102
11.1	Introduction	102
11.2	Conditions	102
11.2.1	Excess noise/vibration	103
11.2.2	Inoperative	103
11.2.3	Corrosion or mechanical damage	103
11.2.4	Obstructed air flow	104

SECTION ONE: AIR CONDITIONING

12.0   **EVAPORATOR FAN (INDOOR FAN, PLENUM FAN, BLOWER OR AIR HANDLER)** **105**
- 12.1   Introduction 105
- 12.2   Conditions 107
  - 12.2.1   Undersized blower or motor 107
  - 12.2.2   Misadjustment of belt or pulley 108
  - 12.2.3   Excess noise/vibration 109
  - 12.2.4   Dirty fan 110
  - 12.2.5   Dirty or missing filters 110
  - 12.2.6   Inoperative 111
  - 12.2.7   Corrosion/mechanical damage 111

13.0   **DUCT SYSTEM** **112**
- 13.1   Supply and return ducts 112
  - 13.1.1   Conditions 112
- 13.2   Duct insulation 124
  - 13.2.1   Conditions 125

STUDY SESSION 5

14.0   **THERMOSTATS** **133**
- 14.1   Introduction 133
- 14.2   Conditions 134
  - 14.2.1   Inoperative 134
  - 14.2.2   Poor location 135
  - 14.2.3   Not level 136
  - 14.2.4   Loose 137
  - 14.2.5   Dirty 137
  - 14.2.6   Damaged 137
  - 14.2.7   Poor adjustment/calibration 138

15.0   **LIFE EXPECTANCY –** **138**

16.0   **EVAPORATIVE COOLERS** **140**
- 16.1   Introduction 140
- 16.2   Conditions 143
  - 16.2.1   Leaking 143
  - 16.2.2   Pump or fan inoperative 143
  - 16.2.3   Rust, mold and mildew 144
  - 16.2.4   No air gap on water supply 144
  - 16.2.5   No water 144
  - 16.2.6   Poor support for pump and water system 145
  - 16.2.7   Louvers obstructed 145
  - 16.2.8   Missing or dirty air filter 145
  - 16.2.9   Cabinet too close to grade 145
  - 16.2.10   Cabinet or ducts not weather-tight 146
  - 16.2.11   Electrical problems 146
  - 16.2.12   Duct problems 146
  - 16.2.13   Excess noise or vibration 147
  - 16.2.14   Clogged pads 147

SECTION ONE: AIR CONDITIONING

17.0	**WHOLE-HOUSE FAN**	147
17.1	Introduction	147
17.2	Conditions	149
17.2.1	Inoperative	149
17.2.2	Inadequate attic venting	150
18.0	**INSPECTION TOOLS**	154
19.0	**INSPECTION CHECKLIST**	155
20.0	**INSPECTION PROCEDURES**	157

FIELD EXERCISE 1

21.0	**BIBLIOGRAPHY**	167
	ANSWERS TO QUICK QUIZZES	168

SECTION ONE: AIR CONDITIONING

▶ 1.0 OBJECTIVES

In this Module, you will learn how **air conditioning** and **heat pump systems** work, and how to identify the common energy sources and heat transfer media used in central air conditioning. We'll talk about the common system configurations. You will learn how to determine the age and capacity of the cooling system, and how to approximate the adequacy of the system.

Not The Last Word

It is not our goal to turn you into a technician or service person. You will not be able to troubleshoot or repair air conditioning and heat pump systems based on what you will learn in this Module. There are many places to go to learn more about air conditioning and heat pumps and we encourage you to continue to expand your knowledge.

By the time you have finished the Module, you will be able to spot the common performance-related problems with systems and their components. You will be able to follow a testing procedure for central air conditioning and heat pump systems.

SECTION ONE: AIR CONDITIONING

Air Conditioning & Heat Pumps
MODULE

STUDY SESSION 1

1. The first Study Session outlines the Scope of air conditioning inspections as set out in the ASHI® Standards of Practice.

Note: **ASHI®** stands for the American Society of Home Inspectors.

This Session also includes some comments on the Standards and a discussion of the basic principles of central air conditioning.

2. Please read the Cooling/Heat Pumps chapter of **The Home Reference Book** before starting this session.

3. At the end of the Study Session you should be able to –

- name the two basic components of the air conditioning system that have to be inspected to meet the Standards
- name the two areas that have to be included in your report
- list three things that are not required in an air conditioning inspection according to the Standards
- describe what happens to Freon in an evaporator coil
- describe what happens to Freon in a condenser coil
- outline the approximate temperatures of the refrigerant in various parts of the air conditioning system
- define in one sentence each the function of the compressor, condenser, evaporator and expansion device

SECTION ONE: AIR CONDITIONING

- define in one sentence **sensible heat**
- define in one sentence **latent heat**
- define the **latent heat of vaporization** in one sentence
- describe in two sentences how air conditioners dehumidify
- describe the relative temperature of the suction and liquid lines on an operating air conditioning system
- list two types of air conditioners other than air-cooled

4. This Study Session may take you roughly one hour.

5. Quick Quiz 1 is included at the end of this Session. Answers may be written in your book.

Key Words:
- *Air conditioning*
- *Refrigerant*
- *Condenser coil*
- *Evaporator coil*
- *Compressor*
- *Expansion device*
- *Sensible heat*
- *Latent heat*
- *Latent heat of vaporization*
- *Dehumidification*
- *Condensate collection*
- *Evaporator fan*
- *Condenser fan*
- *Suction line*
- *Liquid line*
- *Water-cooled systems*
- *Ground-cooled systems*
- *Evaporative coolers*
- *Whole-house fans*

SECTION ONE: AIR CONDITIONING

▶ 2.0 SCOPE AND INTRODUCTION

2.1 SCOPE

THE ASHI® STANDARDS OF PRACTICE

The following are excerpted from the ASHI® Standards of Practice effective January 1, 2000.

2. PURPOSE AND SCOPE

2.1 The purpose of these Standards of Practice is to establish a minimum and uniform standard for private, fee-paid home *inspectors* who are members of the American Society of Home Inspectors. *Home Inspections* performed to these Standards of Practice are intended to provide the client with information regarding the condition of the *systems* and *components* of the home as inspected at the time of the *Home Inspection*.

2.2 The *inspector* shall:

  A. *inspect:*
  1. *readily accessible systems* and components of homes listed in these Standards of Practice.
  2. *installed systems* and *components* of homes listed in these Standards of Practice.

  B. *report:*
  1. on those *systems* and *components* inspected which, in the professional opinion of the *inspector*, are *significantly deficient* or are near the end of their service lives.
  2. a reason why, if not self-evident, the *system* or *component* is *significantly deficient* or near the end of its service life.
  3. the inspector's recommendations to correct or monitor the reported deficiency.
  4. on any *systems* and *components* designated for inspection in these Standards of Practice which were present at the time of the *Home Inspection* but were not inspected and a reason they were not inspected.

2.3 These Standards of Practice are not intended to limit *inspectors* from:

  A. including other inspection services, *systems* or *components* in addition to those required by these Standards of Practice.
  B. specifying repairs, provided the *inspector* is appropriately qualified and willing to do so.
  C. excluding *systems* and *components* from the inspection if requested by the client.

SECTION ONE: AIR CONDITIONING

9. AIR CONDITIONING SYSTEMS

9.1 The *inspector* shall:

  A. *inspect* the *installed* central and through-wall cooling equipment.
  B. *describe:*
   1. the energy source
   2. the cooling method by its distinguishing characteristics.

9.2 The inspector is NOT required to:

  A. *inspect* electronic air filters.
  B. determine cooling supply adequacy or distribution balance.

13. GENERAL LIMITATIONS AND EXCLUSIONS

13.1 General limitations:

  A. Inspections performed in accordance with these Standards of Practice
   1. are not *technically* exhaustive.
   2. will not identify concealed conditions or latent defects.
  B. These Standards of Practice are applicable to buildings with four or fewer dwelling units and their garages or carports.

13.2 General exclusions:

  A. The *inspector* is not required to perform any action or make any determination unless specifically stated in these Standards of Practice, except as may be required by lawful authority.
  B. *Inspectors* are NOT required to determine:
   1. the condition of *systems* or *components* which are not *readily accessible*.
   2. the remaining life of any *system* or *component*.
   3. the strength, adequacy, effectiveness, or efficiency of any *system* or *component*.
   4. the causes of any condition or deficiency.
   5. the methods, materials, or costs of corrections.
   6. future conditions including, but not limited to, failure of *systems* and *components*.
   7. the suitability of the property for any specialized use.
   8. compliance with regulatory requirements (codes, regulations, laws, ordinances, etc.).
   9. the market value of the property or its marketability.
   10. the advisability of the purchase of the property.
   11. the presence of potentially hazardous plants or animals including, but not limited to wood destroying organisms or diseases harmful to humans.
   12. the presence of any environmental hazards including, but not limited to toxins, carcinogens, noise, and contaminants in soil, water and air.
   13. the effectiveness of any *system installed* or methods utilized to control or remove suspected hazardous substances.
   14. the operating costs of *systems* or *components*.

SECTION ONE: AIR CONDITIONING

15. the acoustical properties of any *system* or *component.*

C. *Inspectors* are NOT required to offer:
1. or perform any act or service contrary to law.
2. or perform *engineering* services.
3. or perform work in any trade or any professional service other than *home inspection.*
4. warranties or guarantees of any kind.

D. *Inspectors* are NOT required to operate:
1. any *system* or *component* which is *shut down* or otherwise inoperable.
2. any *system* or *component* which does not respond to *normal* operating controls.
3. shut-off valves.

E. *Inspectors* are NOT required to enter:
1. any area which will, in the opinion of the *inspector*, likely be dangerous to the *inspector* or other persons or damage the property or its *systems* or *components*.
2. The *under-floor crawl spaces* or attics which are not *readily accessible.*

F. *Inspectors* are NOT required to *inspect:*
1. underground items including, but not limited to underground storage tanks or other underground indications of their presence, whether abandoned or active.
2. *systems* or *components* which are not installed.
3. *decorative* items
4. *systems* or *components* located in areas that are not entered in accordance with these Standards of Practice.
5. detached structures other than garages and carports.
6. common elements or common areas in multi-unit housing, such as condominium properties or cooperative housing.

G. *Inspectors* are NOT required to:
1. perform any procedure or operation which will, in the opinion of the *inspector,* likely be dangerous to the *inspector* or other persons or damage the property or its *systems* or *components.*
2. move suspended ceiling tiles, personal property, furniture, equipment, plants, soil, snow, ice, or debris.
3. *dismantle* any *system* or *component,* except as explicitly required by these Standards of Practice.

SECTION ONE: AIR CONDITIONING

GLOSSARY OF ITALICIZED TERMS

Alarm Systems
Warning devices, installed or free-standing, including but not limited to; carbon monoxide detectors, flue gas and other spillage detectors, security equipment, ejector pumps and smoke alarms

Architectural Service
Any practice involving the art and science of building design for construction of any structure or grouping of structures and the use of space within and surrounding the structures or the design for construction, including but not specifically limited to, schematic design, design development, preparation of construction contract documents, and administration of the construction contract

Automatic Safety Controls
Devices designed and installed to protect *systems* and *components* from unsafe conditions

Component
A part of a *system*

Decorative
Ornamental; not required for the operation of the essential *systems* and *components* of a home

Describe
To *report* a *system* or *component* by its type or other observed, significant characteristics to distinguish it from other *systems* or *components*

Dismantle
To take apart or remove any component, device or piece of equipment that would not be taken apart or removed by a homeowner in the course of normal and routine homeowner maintenance

Engineering Service
Any professional service or creative work requiring engineering education, training, and experience and the application of special knowledge of the mathematical, physical and engineering sciences to such professional service or creative work as consultation, investigation, evaluation, planning, design and supervision of construction for the purpose of assuring compliance with the specifications and design, in conjunction with structures, buildings, machines, equipment, works or processes

Further Evaluation
Examination and analysis by a qualified professional, tradesman or service technician beyond that provided by the home inspection

Home Inspection
The process by which an *inspector* visually examines the *readily accessible systems* and *components* of a home and which *describes* those *systems* and *components* in accordance with these Standards of Practice

SECTION ONE: AIR CONDITIONING

Household Appliances
Kitchen, laundry, and similar appliances, whether *installed* or free-standing

Inspect
To examine *readily accessible systems* and *components* of a building in accordance with these Standards of Practice, using *normal operating controls* and opening *readily openable access panels*

Inspector
A person hired to examine any *system* or *component* of a building in accordance with these Standards of Practice

Installed
Attached such that removal requires tools

Normal Operating Controls
Devices such as thermostats, switches or valves intended to be operated by the homeowner

Readily Accessible
Available for visual inspection without requiring moving of personal property, dismantling, destructive measures, or any action which will likely involve risk to persons or property

Readily Openable Access Panel
A panel provided for homeowner inspection and maintenance that is within normal reach, can be removed by one person, and is not sealed in place

Recreational Facilities
Spas, saunas, steam baths, swimming pools, exercise, entertainment, athletic, playground or other similar equipment and associated accessories

Report
To communicate in writing

Representative Number
One *component* per room for multiple similar interior *components* such as windows and electric outlets; one *component* on each side of the building for multiple similar exterior *components*

Roof Drainage Systems
Components used to carry water off a roof and away from a building

Significantly Deficient
Unsafe or not functioning

Shut Down
A state in which a *system* or *component* cannot be operated by *normal operating controls*

SECTION ONE: AIR CONDITIONING

Solid Fuel Burning Appliances
A hearth and fire chamber or similar prepared place in which a fire may be built and which is built in conjunction with a chimney; or a listed assembly of a fire chamber, its chimney and related factory-made parts designed for unit assembly without requiring field construction

Structural Component
A *component* that supports non-variable forces or weights (dead loads) and variable forces or weights (live loads)

System
A combination of interacting or interdependent *components,* assembled to carry out one or more functions

Technically Exhaustive
An investigation that involves dismantling, the extensive use of advanced techniques, measurements, instruments, testing, calculations, or other means

Under-floor Crawl Space
The area within the confines of the foundation and between the ground and the underside of the floor

Unsafe
A condition in a *readily accessible, installed system* or *component* which is judged to be a significant risk of personal injury during normal, day-to-day use. The risk may be due to damage, deterioration, improper installation or a change in accepted residential construction standards

Wiring Methods
Identification of electrical conductors or wires by their general type, such as "non-metallic sheathed cable" ("Romex"), "armored cable" ("bx") or "knob and tube", etc.

▶ NOTES ON THE STANDARDS

Inspect — The Standards are clear on the meaning of **inspect**. When we inspect we have to look at and test the components listed in the Standards. We look at them if they are **readily accessible** or if we can get at them through **readily openable access panels**. These are panels designed for the homeowner to remove. They are within normal reach, can be removed by one person, and are not sealed in place.

SECTION ONE: AIR CONDITIONING

Testing — We test components and systems by using their **normal operating controls**, but not the safety controls. We turn thermostats up or down, open and close doors and windows, turn light switches and water faucets on and off, flush toilets, etc. We do not test heating systems on high limit switches, test pressure relief valves on water heaters and boilers, overload electrical circuits to trip breakers, etc.

Systems Shut Down — We do not start up systems that are shut down. If the furnace pilot is off, we don't light it. If the electricity, water or gas is shut off in the home, we don't turn it on. If the disconnect for the air conditioner is off, we don't turn it on.

Deficiencies — We have to report on systems that are **significantly deficient**. This means they are unsafe or not performing their intended function.

End Of Life — We are required to report on any system or component that in our professional opinion is **near the end of its service life**. This is tricky since we don't know whether inspectors will be held accountable for failed components on the basis that we should have known the component was near the end of its life. With the wisdom of hindsight, it may be hard to argue that the component could not have been expected to fail, when in fact, it did. Time will tell. The situation is also tricky because it includes not only **systems** but individual **components** as well. For many systems there are broadly accepted life expectancy ranges, but these aren't available for some individual components.

Remaining Life — We are not required to determine the **remaining life** of systems or components. This is related to, but different than, the **end of service life** issue. If the item is new or in the middle part of its life, we don't have to predict service life, even though the same broadly accepted life expectancy ranges would apply. It's only when the item is near the end, in your opinion, that you have to report it.

Reporting Implications — We have to tell people in writing the **implications** of conditions or problems unless they are self-evident. A cracked heat exchanger on a furnace has a very different implication for a homeowner than a cracked windowpane, for example.

Tell Client What To Do — We have to tell the client in the report what to do about any conditions we found. We might recommend they repair, replace, service or clean the component. We might advise them to have a specialist further investigate the condition. It's all right to tell the client to monitor a situation, but you can't tell them that their roof shingles are curled and leave it at that.

What We Left Out — We have to report anything that we would usually inspect but didn't. We also have to include in our report why we didn't inspect it. The reasons may be that the component was inaccessible, unsafe to inspect or was shut down. It may also be that the occupant or the client asked you not to inspect it.

Installed Systems Only — The Standards tell us that we have to observe installed air conditioning systems. We do not have to inspect portable air conditioners including window-type systems.

Controls — We have to look at the normal operating controls, which are the thermostat and electrical disconnect.

SECTION ONE: AIR CONDITIONING

Ducts, Coils And Pipes — We have to inspect ducts, registers, grilles and air filters as well as fan coil units and any piping systems. We have to look for insulation on the ductwork and examine how it is supported.

Air To Each Room — We recommend that you make sure there is a source of conditioned air in each room. In many cases there is not, and you will regret having described a house as being air conditioned if the rear addition, for example, has no air conditioning. This is especially true if it is a south-facing solarium with a large glass area.

Air Filters — The air conditioning inspection includes checking the air filters, but not electronic air filters.

Energy Source — We have to describe the **energy source**. In almost all cases that will be electricity. There are a few gas or propane powered central air conditioners.

Cooling Types — We also inspect the **cooling method by its distinguishing characteristics**. You can describe air conditioning systems by where they dump the heat from the house. The most common is an **air-to-air** system where heat is taken out of the house air and discharged into the outdoor air using a refrigerant such as Freon 12 or Freon 22 as the heat transfer medium. We can also dump heat from the house air into water in the house, water outside the house or the ground outside the house.

Piping And Fan Coils — Some systems use water as an intermediate heat transfer medium between the refrigerant and the house air. These systems use piping to move chilled water to a fan coil. The water is passed through the coil and house air is blown over the outside of the coil.

Integral With Heating Or Independent — You can also describe the cooling equipment by its relationship with the heating system ductwork. Many systems use the same ducts, registers and air handlers as the house heating system. Others have independent ducts, registers and fans for the air conditioning system. Through-wall systems have no ducts. They are designed to cool only the space or room they are in, although homeowners often expect more of them.

Evaporative Coolers — Evaporative coolers are a less complicated central cooling system found in hot dry climates, such as Texas, Nevada, Arizona and New Mexico. These systems cool the air coming into the house by passing water droplets through the air.

Test Using Thermostat — The Standards ask us to operate the air conditioning system using the normal operating controls. In most cases, this means turning down the thermostat. There are some limitations where you can't operate the system (when the outdoor temperature is below 65°F, for example).

SECTION ONE: AIR CONDITIONING

Opening Access Panels — The Standards require that you open **readily openable access panels** provided by the manufacturer or installed **for routine homeowner maintenance**. On central air conditioning systems, there are may be no panels intended for opening to facilitate routine homeowner maintenance other than the access to the filters. You are required to open access panels to get a look at the filters, but are not required to open the access panel on the condenser unit or access panels to the evaporator coil. Many home inspectors do open these panels, but this is beyond the Standards.

What We Don't Do
1. The Standards say that we don't have to operate systems when the weather conditions are such that we may damage the air conditioner.
2. We don't have to test portable or window air conditioners.
3. We do not have to determine the adequacy or the uniformity of the air supply to every room.

Capacity — The Standards don't ask us to observe or describe the capacity of the system, although many inspectors do.

Rare Systems — We won't discuss in detail the rare air conditioning systems, including **gas chillers** (absorption cooling) and systems that use piped cool water and fan coils.

2.2 INTRODUCTION

Luxury — Central air conditioning is considered a luxury rather than an essential in most parts of North America. Air conditioning systems are more common in the southern areas, and are more common in humid areas than dry areas. There are probably more air conditioners per capita in Florida than in California, for example. You probably have a good sense already as to how common central air conditioning systems are in your area.

We inspect air conditioning systems that are connected to ductwork, and which deliver conditioned air to all parts of the house. The Standards also ask us to inspect through-the-wall and other ductless systems.

Through-the-wall Systems — Through-the-wall air conditioners do not have a duct system. These units are commonly used to heat single rooms or areas without partition walls. They may be self-contained units or may be split-systems with the evaporator and fan in the wall, and a remote compressor condenser coil and fan.

Split Systems — Split-system central air conditioners are the most common. They can be independent systems or incorporated with indoor furnaces and indoor fan coils that can be upflow, downflow or horizontal. You may also come across single-package systems that are often on the roof, in attics or in crawlspaces. The supply and return duct system extends outside the living space to pass the air through the air conditioning or heat pump system.

SECTION ONE: AIR CONDITIONING

Many Limitations

As with all systems, you can't inspect everything, and with air conditioning systems, there are many limitations. For example, you shouldn't test an air conditioner when the outdoor temperature is below 65°F or has been below 65°F in the last 24 hours. If the power to the central air conditioner has been turned off, you cannot simply turn the power on and test the air conditioner, as this may damage the compressor. In many cases, you can't get a good look at the internal components of the condenser unit (removing the access panel to the condenser unit is beyond the Standards) and, sometimes, you won't get a look at the evaporator coil in the ductwork inside the house.

Client Expectations

You may be tempted to think that this isn't important, since the air conditioner is a luxury item only, and the house will still be habitable even if the air conditioner doesn't work. However, you should understand that the clients bought the house with central air conditioning and expect it to work. If it doesn't work and you didn't warn them of this, you should expect to receive a callback from the client.

It's worth noting that central air conditioning systems are among the most expensive mechanical components in a home and have a relatively short life expectancy.

SECTION ONE: AIR CONDITIONING

▶ 3.0 THE BASICS

The challenge

Complicated Air conditioning is complex. To most people, it is not obvious how on a hot summer day, you can take heat out of a house and throw it outside, where it is even hotter. But that is exactly what central air conditioning does.

We'll use the split-system air-cooled central air conditioning system in this discussion since it's the most common system in homes.

More Than Cooling Comfort involves more than cool air. Air conditioning also involves more than lowering the air temperature. It includes dehumidifying, cleaning (filtering) and circulating the air. Good air conditioning systems perform all of these functions, although most people focus on the "cool" concept. (In the broadest sense of the term, **air conditioning** also means heating, humidification and ventilation, although we'll exclude these issues from our discussion.)

House Slowly Heats Up If the outdoor temperature is 70°F at night, and all the windows in the house are open, the indoor and outdoor temperatures will both be about 70°F. As the sun comes up, the outdoor temperature may rise to 85°F or 90°F. Because of shading, thermal mass and so on, the house will not heat up as quickly as the outdoors, but it will eventually get just as hot as it is outside. The goal is to keep it more comfortable inside the house than it is outside.

The mechanics

The most common type of air conditioning that we see is technically referred to as **direct expansion, mechanical, vapor-compression refrigeration system**.

Like A Refrigerator The goal with air conditioning is to capture heat in the house and throw it outside. But how can we take heat from a space that is already cooler than outdoors and dump it into the outdoor air? One of the ways we can think about it is to look at a refrigerator. If we can keep the temperature inside your refrigerator at about 40°F and it is 70°F in the kitchen, somehow we are taking heat out of that cool air and dumping it into a kitchen that is warmer. Central air conditioning and refrigerators operate on exactly the same principle. The process works something like this.

SECTION ONE: AIR CONDITIONING

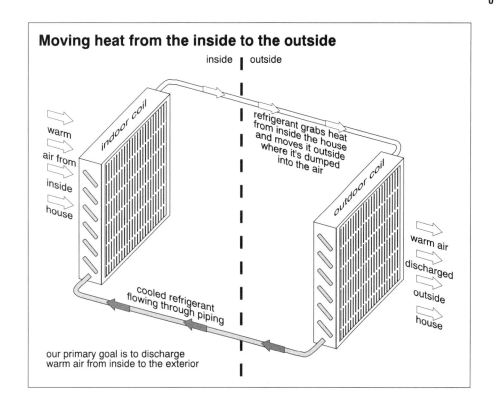

The Coils   We have two coils similar to the radiator in a car: one inside the house and one outside. We put something cold through the coil inside the house and then blow warm house air across the coil, so the coil can grab heat from the house air. This cools the house. We want to take that heat in the coil outside and dump it into the outdoor air. Let's look at how we can do this.

Freon   On the inside of a coil we use a substance such as **Freon 12** or **Freon 22** (which are brand names for a **refrigerant** that is non-corrosive, non-flammable, non-toxic, but as we have recently discovered, not great for the ozone layer). This refrigerant is a colorless gas at atmospheric temperature and pressure. Inside the coils we manipulate the Freon to make it a liquid or a gas.

SECTION ONE: AIR CONDITIONING

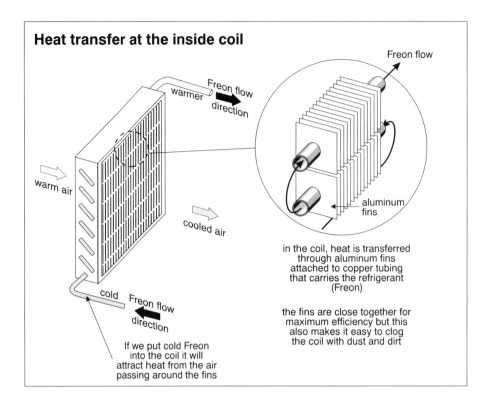

Freon Goes In Circles

The Freon runs in a loop, passing through the indoor coil, through a copper pipe to the outdoors, through the outdoor coil and back inside through another pipe to the indoor coil.

Cool Freon Liquid Into Evaporator

Let's follow the Freon from the point that it comes into the evaporator (indoor) coil. As it enters the coil, it is a cold liquid, perhaps 20° to 40°F. The cold liquid in the coil feels the warm house air on the other side of the coil. If the house air is about 75°F and the Freon is at 30°F, heat is going to move through the coil (which is just a heat exchanger) and warm the liquid. As it warms up, the liquid boils off into a gas.

Freon Sucks Heat From House

As the Freon inside the coil changes from a liquid to a gas, it sucks heat out of the house air. Logically enough, this is called the **evaporator coil** because the Freon inside is evaporating from a liquid to a gas.

50°F Gas Leaves Evaporator

The Freon leaves the evaporator coil as a gas that is warmer than the liquid coming in, but still cooler than the air around it. The temperature of the gas might be 50°F.

SECTION ONE: AIR CONDITIONING

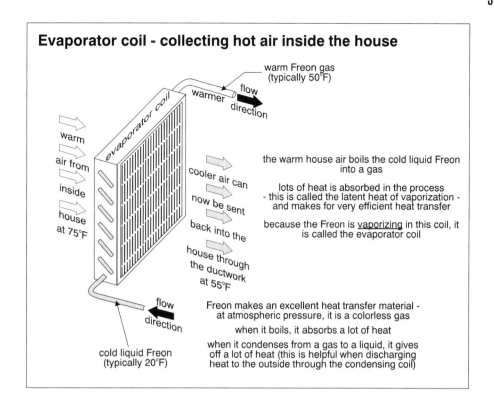

Gas Goes Outside

Now we want to dump the heat from the Freon gas outdoors. We have a problem. If you take 50°F Freon and pass it through an outdoor coil where the air temperature is 85°F or 90°F, we are just going to heat up that Freon gas and actually collect more heat. That won't work! What we want to do is get rid of the heat.

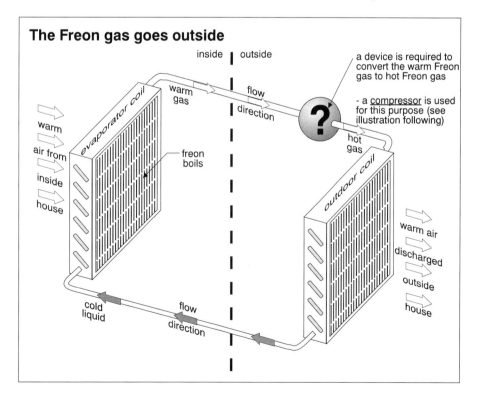

SECTION ONE: AIR CONDITIONING

Compress The Gas To Heat It Up

The solution involves a **compressor,** which we can think of as a pump. This compressor squeezes the gas, which heats it up. There is a gas law that says that if you increase the pressure on a gas, you also increase its temperature. This is great! We take a gas at 50°F, squeeze it really hard to build up pressure and raise the temperature. The low pressure gas at 50°F that entered the compressor, leaves the compressor at a much higher pressure and **temperature.**

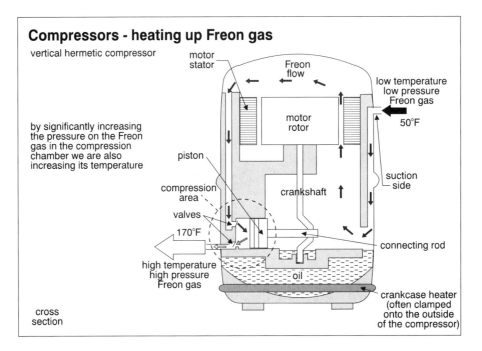

Hot Gas To Condenser Coil

If the temperature coming out is 170° (to 230°F), we can pass this hot gas through the outdoor coil and blow some outdoor air across it. The Freon is now able to dump its heat into the 85°F or 90°F outdoor air. So, the heat we removed from the house can be thrown away outdoors! Before you go any further, make sure you understand the mechanics so far.

Gas Cools And Condenses

As the hot, high pressure gas moves through the outdoor coil and gives off its heat, it cools to the point where it condenses back to a liquid. Logically enough, the outdoor coil is called the **condenser** coil because the Freon inside condenses from a gas to a liquid.

SECTION ONE: AIR CONDITIONING

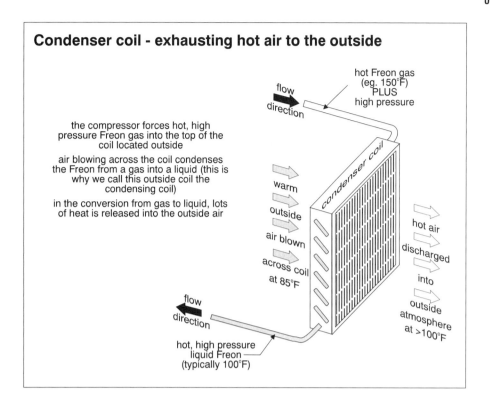

Incidentally, an air conditioner is working properly if the air coming off the outdoor fan is even hotter than the outdoor air. That is because we are passing the 85°F or 90°F air across a coil where the Freon gas inside is at 170°F.

Getting Ready To Boil The Freon Again

After the Freon goes through the condenser coil, we end up with a high pressure liquid that is still relatively warm. It might be between 95°F and 110°F, for example. The compressor is pushing this hot, high pressure liquid through a pipe back into the house.

Hot Liquid Back To House

This liquid is too warm to pick up heat from the house. At 95°F to 110°F, this is not going to allow us to steal any more heat from the house. Now we have another problem.

SECTION ONE: AIR CONDITIONING

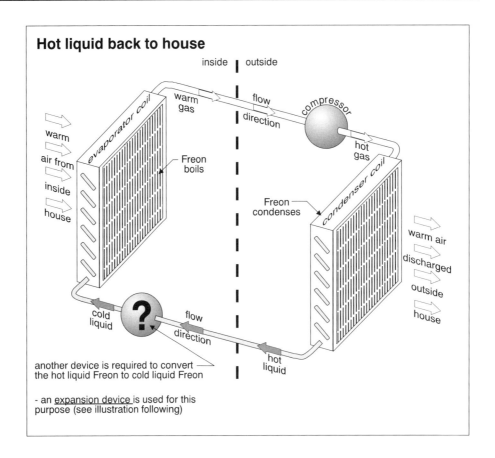

Lower The Pressure To Cool The Liquid

However, we can use another little trick to cool off that liquid. If we pass it through a restriction (a **capillary tube** or **thermostatic expansion valve**, for example), we can allow only a little bit of the high pressure liquid to move through the pipe at a time. This means that on the discharge of the bottleneck, the liquid will be at a much lower pressure.

SECTION ONE: AIR CONDITIONING

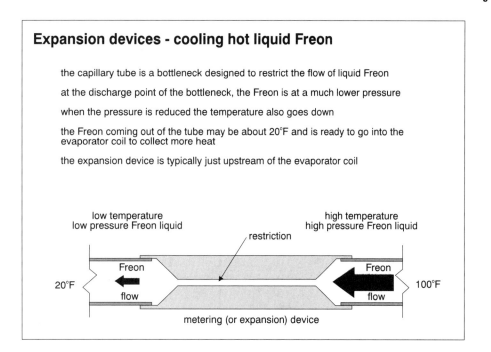

Cool Liquid To Evaporate As we lower the pressure, we also the lower the temperature. The liquid that came into the bottleneck at 95°F to 110°F comes out colder (20° to 40°F) and may already be starting to boil! Now we are ready to go through the evaporator coil again, collecting heat from the 75°F house air and boiling the Freon off to a gas.

Review

Indoors To recap, we can think about a cold, low pressure liquid entering an evaporator coil. The warm house air gets blown across the cool coil by the furnace fan. The house air gives up its heat to the cold liquid, boiling the liquid off into a relatively cool gas. The cooled air which passed over the coil is distributed through the house.

Outdoors The cool gas in the pipe moves outside and is squeezed by the compressor into a high temperature, high pressure gas. This hot gas passes through the condenser coil. Blowing outside air across the condenser cools the hot gas inside, releasing heat to the outdoor air that has been stolen from the house. As the gas is cooled, it condenses back to a liquid.

Indoors Again This hot liquid gets pushed back into the house. Just before it goes into the evaporator coil, it goes through the bottleneck (capillary tube or expansion valve) which lowers the pressure and temperature dramatically. Again, we have the cold liquid entering the evaporator coil, starting the process over.

SECTION ONE: AIR CONDITIONING

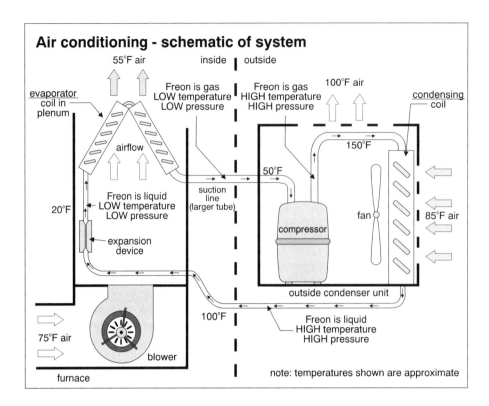

Sponge Analogy	You can think of air conditioning as a sponge. The sponge contains heat from the house. We squeeze (compress) it to get rid of the water (heat) outside then let it expand (expansion device) as it comes inside so we can soak up more water (heat).

Comfort — what causes it?

Sensible Heat	We should define a couple of terms here. **Sensible heat** is the heat a thermostat senses. When the temperature goes up, there has been an increase in the sensible heat.
Latent Heat	**Latent heat** is hidden. It involves adding or removing heat without changing the temperature. How does this happen? When we change a liquid to a gas (boiling or evaporation), we have to add heat, but we don't need to change the temperature. Boiling 212°F water produces 212°F steam. Similarly, we can remove heat by condensing gases to liquids, without lowering the temperature. Air contains latent heat in the water vapor that is in the air. Removing the vapor removes heat, but doesn't lower the temperature.
Dry Air Feels Cooler	Let's think about comfort. It is easy to understand that people are more comfortable at 70°F than 100°F. But there is more to it. Most people have experienced how much more uncomfortable it is on a hot, humid day than on a hot, dry day. It helps to understand why that is.
Evaporative Cooling	The human body cools itself by sweating. When moisture evaporates off the surface of the skin, there is a great deal of cooling that takes place. Dogs accomplish much the same thing by panting.

SECTION ONE: AIR CONDITIONING

There is something called the **latent heat of vaporization**. The key is that it takes lots of energy to convert liquid to a gas.

Latent Heat Of Vaporization

Here is an example of the latent heat of vaporization. One **British Thermal Unit (BTU)** is defined as the amount of heat required to raise one pound of water one Fahrenheit degree. To heat one pound of water from 32°F to 212°F, you have to add 180 BTUs to it (212° - 32°F = 180°F). To turn that one pound of 212°F water into steam (which is also 212°F), you have to add another **970 BTUs** to it! It takes five times as much energy (heat) to boil water than it takes to warm it up from 32° to 212°F. Notice that the steam is the same temperature as the water despite the addition of all that heat!

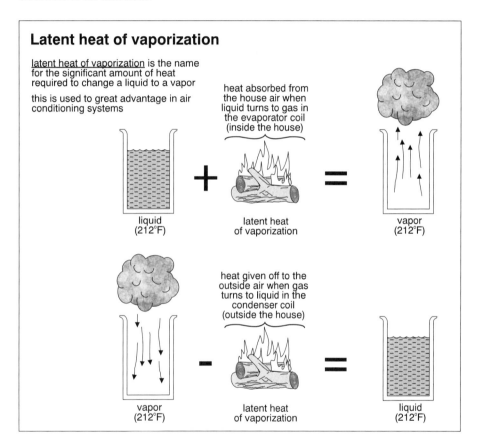

When a liquid changes to a gas, a tremendous amount of energy is absorbed. That's why having sweat evaporate off our skin is so helpful in keeping us cool.

Low Humidity Lets Skin Moisture Evaporate

People are comfortable when the humidity is lower because it is easier for the moisture on their skin to evaporate. The process of evaporation removes heat. It is also easy to understand how fans help keep people comfortable. As water evaporates off the skin, the air immediately adjacent to the skin becomes saturated and cannot hold any more moisture. The faster that air moves across your skin, the more quickly the saturated air is carried away and replaced by dry air, which allows more evaporation.

Lower Humidity More Comfortable In Summer

So if we want to keep people comfortable inside their houses, we need to cool the air and lower the humidity. If we can keep the house about 15°F to 20°F cooler than it is outside, that is usually adequate. In the winter, we set our thermostats around 70°F. In the summer, 75°F is usually just fine. If the outdoor temperature reaches 100°F, we cannot expect an air conditioning system to keep the house at 75°F. But if it is that hot outside, even 80°F with low humidity will feel relatively comfortable inside the house.

The latent heat of vaporization is also useful in the Freon lines. We can collect heat when boiling Freon and release heat when condensing it. It helps make air conditioners efficient.

Dehumidifying

Air Conditioners Dehumidify

We have explained how the house temperature can be dropped in an air conditioner but have not addressed the humidity. Part of the function of central air conditioning is to dehumidify the house air. Let's look at humidity reduction. When you pour lemonade into a glass on a hot summer day, the outside of the glass may sweat. That's because you are cooling the air around the glass down to a point where it cannot hold all of the moisture that's in it as vapor, and the humidity falls out as condensation.

Sweat Collects On Evaporator Coil

The evaporator coil works the same way. If you think of the evaporator coil as a cold glass of lemonade, it is easy to imagine how the outside of that coil will sweat as the warm, moist air from the house passes over it. The liquid in the coil is very cold. As the air outside the coil gives off heat, it also loses its ability to hold moisture. Condensation forms on the outside of the evaporator coil. The air gives up its moisture as it passes over the coil. As we discussed, the dryer air helps people feel cooler by allowing sweat to evaporate from their skin.

Condensate Is Collected

The moisture that forms on the coil is collected in a pan below the coil. The water is drained from the pan through a condensate tube to a floor drain or sink, to the outside or some other acceptable discharge point.

SECTION ONE: AIR CONDITIONING

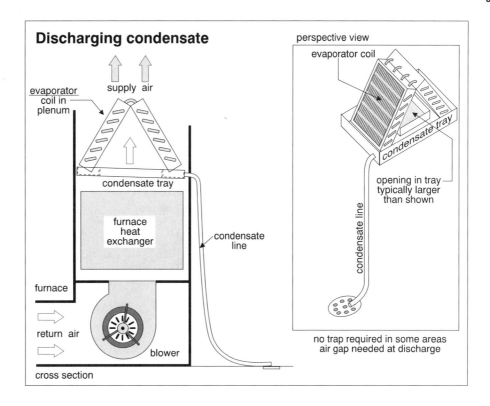

System Components

Fans And Coils

When an air conditioning system is running, fans are blowing air across the evaporator and condenser coils. Inside the house, we use the furnace fan if there is one. If there is no forced-air heating, we put in a separate fan to move air across the coil and through the ducts. Outdoors we put a fan in the cabinet with the condenser and compressor.

The evaporator coil and condenser coil in an air conditioner are heat exchangers although they are not normally referred to as such. Their function is to transfer heat from the house air into the refrigerant inside and to move heat from the refrigerant into the outdoor air.

Refrigerant

The refrigerant is the vehicle that collects from the house, moves it outside and releases it into the outdoor air.

Compressor And Expansion Device

The compressor in the condenser cabinet is squeezing a cool, low-pressure gas into a hot, high-pressure gas. The expansion device (capillary tube or thermostatic expansion valve) near the evaporator coil is converting a hot, high-pressure liquid to a cool, low-pressure liquid.

SECTION ONE: AIR CONDITIONING

The Freon lines

High-Pressure And Low-Pressure Sides

Some people talk about the **high-pressure** and **low-pressure** sides of an air conditioning system. The high-pressure side is from the discharge side of the compressor, through the condenser coil, through the liquid line and up to the expansion device. The low-pressure side is from the expansion device, through the evaporator coil, and out through the suction line to the inlet side of the compressor.

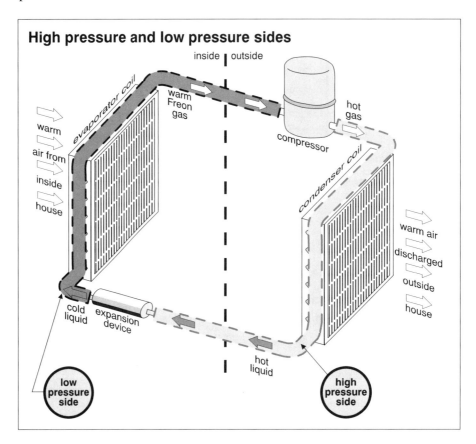

Suction Line Insulated

The larger copper tube (suction line) that carries the cool low-pressure gas from the evaporator coil out to the compressor is insulated. This is because we do not want to dump more heat into the gas as we move outside. Remember, we are trying to dump heat from inside the house to the outdoors, not collect outdoor heat.

Gas Line Should Be Cool

The other reason to insulate the suction line is to prevent sweating on the pipe. This gas is cool, relative to the house air and the outside air. If un-insulated, it would have condensation all over it.

When you are inspecting and operating air conditioners, you should check to see that the large suction line is cool to the touch any place where there is no insulation. If there is condensation on the small un-insulated sections of the pipe, that is fine. You do not want to see frost; that indicates a problem.

SECTION ONE: AIR CONDITIONING

Liquid Line Should Be Warm

The un-insulated copper line coming from the condenser coil back into the house contains a warm liquid. It is smaller than the suction line which carries a gas. This is logical because gas takes up more space than liquid. Because this smaller line carries a warm liquid, when the air conditioner has been running for fifteen minutes or so, this line should feel warm to the touch.

Is it working properly?

When an air conditioner is running properly at steady state, the house air temperature drops by 15°F to 20°F (some say 14°F to 22°F) as it moves across the evaporator coil. For example, 75°F house air would come out of the evaporator coil at 55°F to 60°F.

One can also make sure that the air coming off the condenser fan outside is warmer than the outside air. You can sense this with your hand.

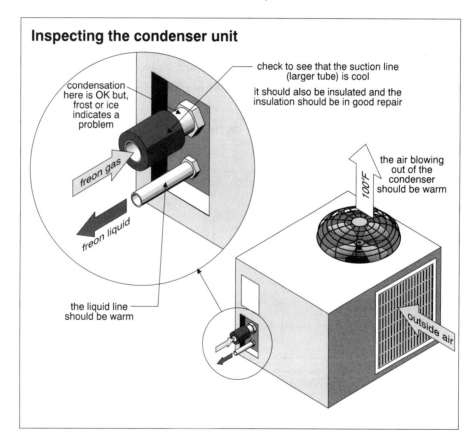

Seasonal Energy Efficiency Ratio

The SEER is simply a ratio of how many BTUs per hour you're getting out of the system relative to the watts of electrical energy consumed to run the unit.

SEER = $\dfrac{\text{Total Cooling Output over Season}}{\text{Total Electrical Input over Season}}$

SECTION ONE: AIR CONDITIONING

SEER ratings of 6 are typical for old air conditioners. New air conditioners are typically around 10, and high-efficiency air conditioners are typically about 14. While some clients may ask, most inspectors do not report on efficiencies, and the Standards don't require it.

Summary

To summarize the process one more time, we take a cold, low pressure liquid and pass it across the evaporator coil inside the house. The 75°F house air blows across it and comes off the coil at 55°F or 60°F. The cold liquid Freon that went into the evaporator coil comes out as a low pressure gas.

This cool, low pressure gas is taken outside and compressed into a hot, high pressure gas. This hot gas is passed through the condenser coil where 85°F to 90°F (for example) outdoor air is blown across it. The hot gas gives off some of its heat to the outdoor air. This causes the hot gas to condense back to a warm liquid. The outdoor air may enter the coil at 85°F and leave at 100°F (a 15°F to 20°F temperature rise is typical.) The warm liquid is carried back into the house, where it passes through the expansion device (bottleneck). This drops its temperature and pressure so that it can enter the evaporator coil as a cold liquid, ready to get boiled off.

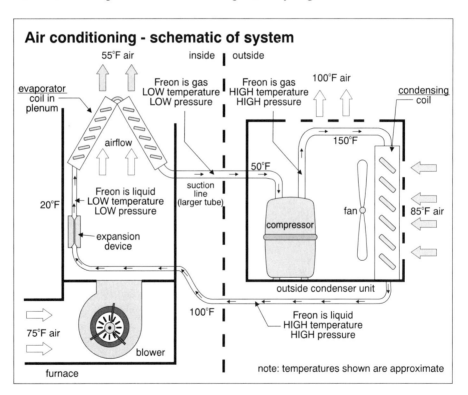

One little twist

Water-Cooled Systems

Some people are not happy keeping things simple. They decided if we could dump heat into the outside air, we could also dump it into water. So they built air conditioners that dump heat into the domestic water in the house. Since this water is usually about 50°F, we can dump a lot of house heat into it.

SECTION ONE: AIR CONDITIONING

Water Passes Over Condenser

The evaporator coil sits in the house air system just like we have been talking about, and works the same way as before. And we still use a condenser coil. There is Freon inside the coil (as usual), but on the outside is water. The water flows through the condenser coil without the need for a fan. The water is pushed by city water pressure (or pump pressure on a private water supply).

Condenser And Compressor Are Indoors

The condenser coil does not have to go outside any more. We can put it close to the evaporator coil. That means the compressor can be indoors too. On the downside, the water that passes through the air conditioning system cannot be used for drinking or washing. It is no longer considered **potable.**

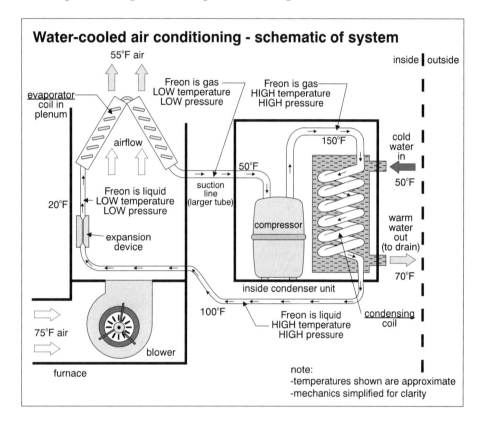

Waste Water

Some maintain that these air conditioning systems are wasteful, since the outgoing water is usually dumped down a drain. Some systems can use the air conditioning discharge water to water the lawn or fill a swimming pool, but even then, some municipalities have banned these systems because of their high water consumption.

Rivers, Lakes, Wells And Ground

There are other places to dump the heat, including lakes, rivers, ponds, wells, and the ground. Systems have been developed to use these. All of these systems operate on the same principle we have discussed. Many use an intermediate liquid (brine, anti-freeze, etc.) to carry heat from the condenser to the water or ground. These systems typically use pumps to move the intermediate liquid.

SECTION ONE: AIR CONDITIONING

Other twists

Gas Chillers Gas chillers work differently than conventional residential air conditioners and have become so rare that they do not warrant discussion here. A brief description can be found in Section 1.4 of the Cooling/Heat Pumps chapter of **The Home Reference Book**.

Evaporative Coolers We'll briefly discuss evaporative coolers, an entirely different air conditioning system used in very dry climates.

Whole House Fans We'll also touch on whole-house fans, which aren't really air conditioners at all, but do help to cool houses.

Heat Pumps Heat pumps will be discussed in the next section. We strongly encourage you to finish the air conditioning discussion before starting the heat pump section.

Ductless Air Conditioning

Alternative To Central Systems Air conditioning is expensive to add to homes that do not have ducts. **Ductless air conditioning** is becoming a popular option because it is not disruptive to install and can provide several cooling zones. Ductless systems are available as air- or water-cooled units.

Scope Of Inspection We inspect permanently **installed** systems with or without ducts. We don't inspect window air conditioners or other systems that can be removed without tools and that plug into 120-volt convenience receptacles.

Split Systems There are two common types of ductless systems; **split systems** and **single component systems**. Split systems have a condenser cabinet with a compressor, condenser coil and fan on the ground or on the roof, the same as any central split system. The evaporator coil and house air fan are inside the home, in the area to be cooled. There is a condensate collection and discharge system for the interior component. There are two refrigerant lines, often in a conduit, joining the outdoor condenser unit to the evaporator inside the home.

Compact Split systems, also called **mini-splits** are easy to install and only require a 3-inch diameter hole through the house wall. The indoor components can be wall or ceiling mounted and don't take up much space. Some are sold with remote controls so they can be mounted out of the way, high on walls or on ceilings.

Quiet These have the advantage of a remote compressor (the noisiest part of an air conditioner) so the home is quieter. Some interior fans are multi-speed to minimize noise. There are also quieter condenser fans in some systems that operate at very low rpm (less than 900 rpm).

Multi-zone Systems Split systems can be multi-zone, with one condenser unit serving up to four evaporators in four different parts of the home.

SECTION ONE: AIR CONDITIONING

Large Capacity
Split systems are available with cooling capacities up to 60,000 BTUs/hr (five tons).

Heat Pumps Too
Heat pumps are also available as ductless sytems. Some have built-in supplementary electric heating.

Single Component Systems
Single component systems are also called **through-wall** or **package systems**. These are self-contained systems with the condenser, compressor and evaporator all in the same cabinet, installed in the wall of the room or area to be cooled. These units are common in motels and apartments. These single component systems are noisier than split systems because the compressor is in the wall. Some include electric elements for supplementary heating. Single component systems may be wired directly into the panel or may plug into a 240-volt receptacle.

Inspection Issues
The inspection procedures for ductless systems are similar to central systems. There is no distribution system to worry about, and there is often much less you can see, especially on single component systems. There is typically a filter access panel, but that is about all that is accessible. On ceiling and high-wall mounted units, the Standards suggest you don't have to open these panels since they are not within **normal reach**. Dirty air filters are a common problem with ductless systems, especially when the system is out of reach of the average person.

Airflow Issues
Ductless air conditioners can blow air up to 40 feet in an open area, but since there is no distribution system, even cooling or heating in multiple rooms from a single system is unlikely. In small rooms, air can bounce off walls or furnishings and create short cycling and comfort problems. These systems are often located near the top of the stairwell in a two-story home in an effort to cool as much of the home as possible.

Condensate Damage
Condensate discharge systems are often on the building exterior, below the wall-mounted evaporator. Discoloration or damage to the wall is a possibility if the condensate is allowed to run down the wall surface.

SECTION ONE: AIR CONDITIONING

Air Conditioning & Heat Pumps
MODULE

QUICK QUIZ 1

☑ INSTRUCTIONS

- You should finish Study Session 1 before doing this Quiz.
- Write your answers in the spaces provided.
- Check your answers against ours at the end of this Section.
- If you have trouble with the Quiz, reread the Study Session and try the Quiz again.
- If you did well, it's time for Study Session 2.

1. An inspection done to the Standards requires us to look at the (two correct answers) –

    a. cooling and air handling equipment
    b. normal operating controls
    c. window air conditioners
    d. uniformity of cool air supply
    e. performance of the system in any weather

2. If the air conditioning system has been shut down, then we have to activate it to test it.
    True ☐   False ☐

3. We have to report on the presence of a cool air supply to each room.
    True ☐   False ☐

SECTION ONE: AIR CONDITIONING

4. We have to check filters in the air handling system.
   True ☐   False ☐

5. What is the most common energy source for central air conditioning systems?
   _____

6. To test the air conditioning system, we have to operate the thermostat.
   True ☐   False ☐

7. Central air conditioners are most like which household appliance?
   a. a stove
   b. a refrigerator
   c. a microwave
   d. a trash compactor
   e. a garbage disposal

8. In what state is Freon when it is in the suction line?
   _____

9. What is the temperature of Freon when it leaves the evaporator coil?
   _____

10. What is the temperature of Freon as it enters the evaporator coil?
    _____

11. What is the temperature of Freon as it enters the compressor?
    _____

12. What is the temperature of Freon when it leaves the compressor?
    _____

13. What is the temperature of Freon when it enters the condenser coil?
    _____

14. What is the temperature of Freon when it leaves the condenser coil?
    _____

15. What is the temperature of Freon when it enters the expansion device?
    _____

SECTION ONE: AIR CONDITIONING

16. What is the temperature of Freon when it leaves the expansion device?

17. In what state is Freon in the compressor?

18. In what state is Freon in the expansion device?

19. When you compress gas, you cool it.

True ☐   False ☐

20. The evaporator coil is outdoors in a split system.

True ☐   False ☐

21. Which of the Freon lines is insulated?

22. Define, in one sentence, sensible heat.

23. Define, in one sentence, latent heat.

24. How many BTUs are required to convert one pound of water at 212°F to steam at 212°F?

25. Explain in two or three sentences how people's bodies are cooled by evaporative cooling.

26. Explain in two sentences how air conditioners dehumidify houses.

27. Where is condensate collected in an air conditioning system?

SECTION ONE: AIR CONDITIONING

28. Which side of the air conditioning loop is the high pressure side?

_____

29. Which side of the air conditioning loop is the low pressure side?

_____

30. When the system is operating, which of the refrigerant lines will feel warm?

_____

31. When the system is operating, which of the refrigerant lines will feel cool?

_____

If you didn't have any difficulty with the Quiz, then you are ready for Study Session 2.

Key Words:
- *Air conditioning*
- *Refrigerant*
- *Condenser coil*
- *Evaporator coil*
- *Compressor*
- *Expansion device*
- *Sensible heat*
- *Latent heat*
- *Latent heat of vaporization*
- *Dehumidification*
- *Condensate collection*
- *Evaporator fan*
- *Condenser fan*
- *Suction line*
- *Liquid line*
- *Water cooled systems*
- *Ground cooled systems*
- *Evaporative coolers*
- *Whole house fans*

SECTION ONE: AIR CONDITIONING

Air Conditioning & Heat Pumps
MODULE

STUDY SESSION 2

1. You should have finished Study Session 1 and Quick Quiz 1 before starting this Study Session.

2. This Study Session deals with the capacity of air conditioning systems and the compressor, which is the heart of an air conditioning system.

3. At the end of this Study Session, you should be able to –

   - define what a ton of air conditioning is in btus per hour
   - list ten factors that affect how much air conditioning is needed
   - provide a range of how many square feet can be cooled by a ton of air conditioning in southern and northern climates
   - explain in one sentence the implications of an undersized air conditioning system
   - explain in one sentence the implications of an oversized air conditioning system
   - describe, in one sentence each, the function and location of the air conditioning compressor
   - describe in one sentence the function of a crankcase heater
   - list nine common compressor problems

4. This Study Session should take roughly one hour to complete.

5. Quick Quiz 2 is included at the end of this Session. Answers may be written in your book.

SECTION ONE: AIR CONDITIONING

Key Words:
- ***One ton***
- ***Ductwork capacity***
- ***Undersized***
- ***Oversized***
- ***Temperature drop***
- ***Compressor***
- ***Crankcase heater***
- ***Noise/vibration***
- ***Short cycling***
- ***Out of level***
- ***Excess current***
- ***Wrong breakers***
- ***Undersized wire***
- ***No electrical disconnect***
- ***Inoperative***
- ***Inadequate cooling***
- ***Slugging***
- ***Rated Load Amperage (RLA)***
- ***Full Load Amperage (FLA)***

SECTION ONE: AIR CONDITIONING

▶ 4.0 AIR CONDITIONING CAPACITY

Most home inspectors give their clients some indication as to whether the air conditioning system is sized properly even though it's not required by the Standards. Let's look at the cooling capacity.

4.1 INTRODUCTION

A Ton Of Ice

As discussed in **The Home Reference Book**, the cooling capacity of air conditioners is usually measured in tons. One ton equals 12,000 BTUs (British Thermal Units) per hour. The term **one ton** comes from the amount of heat required to melt a block of ice that weighs one ton.

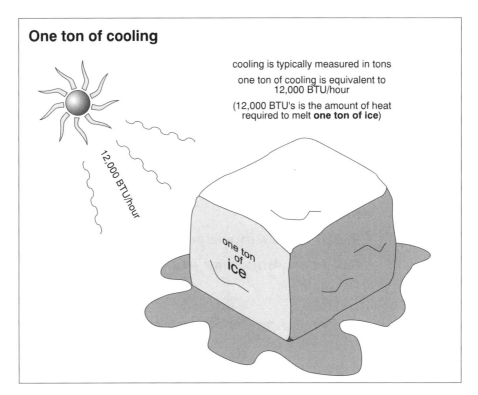

Factors Affecting Cooling Load

The amount of cooling required depends on a large number of factors. These include the outdoor temperature; the outdoor humidity; the level of insulation in the house; the amount of air leakage in the house; the amount of southern, east and west facing glass in the house; whether this glass is single-, double- or triple-glazed; whether the glass is a low emissivity glass; and whether window treatments (curtains or blinds) are kept closed or open. Other factors include the amount of shading from trees, roof overhang, awnings or buildings and how much heat is generated in the house by the people and equipment inside.

SECTION ONE: AIR CONDITIONING

Guidelines Despite all these variables, most people like to have guidelines. Home inspectors are no exception. In the southern United States, 450 to 700 square feet of floor area per ton of cooling is considered appropriate. In the more moderate climates, such as the northern United States and southern Canada, 700 to 1,000 square feet per ton may be adequate. Speak to air conditioning contractors and other inspectors in your area to find the appropriate range for your area. (Note: These guidelines assume 8 foot ceilings.)

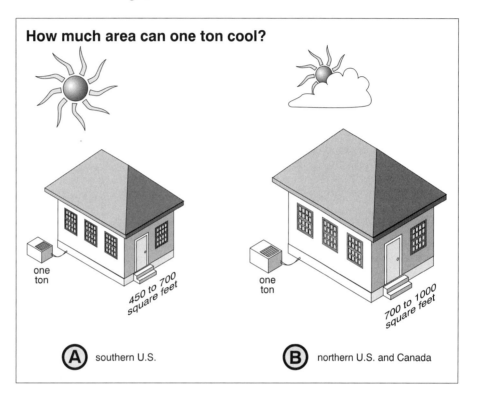

Duct Capacity Problems The capacity of the equipment is only one part of the equation. Many air conditioners that under-perform are a result of a duct system incapable of circulating the conditioned air adequately through the system. This is particularly true where air conditioning has been added to a house with ducts that were designed for a heating system only.

Moving Heavy Air With More Obstructions Adding central air conditioning to an existing furnace system may lead to inadequate air distribution for several reasons. Firstly, the evaporator coil presents an additional obstruction to airflow and reduces the rate of air movement through the system. Secondly, during the cooling season, we are trying to move air that is at 55°F rather than air that is at 140°F (which is what we see in the heating season). The cooler air is more dense (heavier) and more difficult to move through the ducts. We also have to move more air since the difference in temperature between the conditioned air and the room air (about 15°F to 20°F) is less than with a conventional oil or gas furnace (60°F to 70°F), for example.

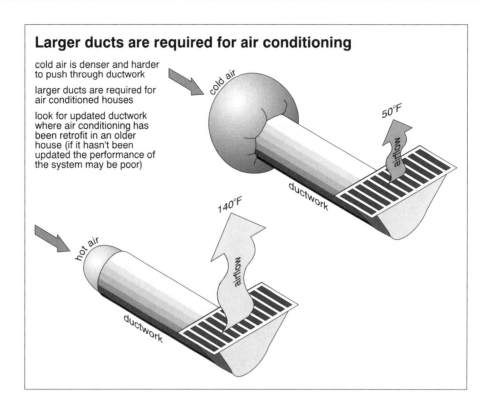

Larger ducts are required for air conditioning

cold air is denser and harder to push through ductwork

larger ducts are required for air conditioned houses

look for updated ductwork where air conditioning has been retrofit in an older house (if it hasn't been updated the performance of the system may be poor)

A larger fan is only helpful up to a point. We don't want to increase the air speed beyond 500 feet per minute (about 5 miles per hour) or we'll get excessive noise and uncomfortable drafts in the home.

Air conditioning systems typically move 400 to 450 cubic feet of air per minute per ton through a system. Heating systems only need to move about half this much.

What Constitutes Good Performance

Most air conditioning systems are designed with a slightly different goal than heating systems. During the heating season, our goal is typically to keep the house at roughly 70°F regardless of how cold it is outside. During the cooling season, while it may be ideal to drop the temperature to 75°F, remember that the air conditioning system is also dehumidifying the air. As long as a 15°F differential between the outdoor temperature and indoor temperature is achieved, the house will feel relatively comfortable if the air has been dehumidified properly. When it's 100°F outside, an indoor temperature of 80°F to 85°F may be acceptable. Clients should understand that this temperature differential indicates good performance.

Uneven Cooling

One of the common complaints with air conditioning is that different levels of the house are cooled with different effectiveness. For example, it may be 75°F on the main floor, but 85°F on the second floor. This is usually a function of the distribution system rather than the capacity of the unit.

SECTION ONE: AIR CONDITIONING

Better To Undersize Than Oversize
Many air conditioning manufacturers and installers recommend slightly undersizing an air conditioning system, rather than oversizing. The reason for this is twofold. First, air conditioners that are slightly undersized tend to have longer running periods. This means fewer stops and starts, and potentially a longer compressor life.

House Is Cold And Damp
Second, and perhaps more importantly, the risk in oversizing a unit is an uncomfortable climate. Oversized air conditioners come on for short periods of time and drop the air temperature quickly. Because of their large capacity, they satisfy the thermostat before the system has a chance to do much dehumidification. This can lead to a cold clammy environment inside the house.

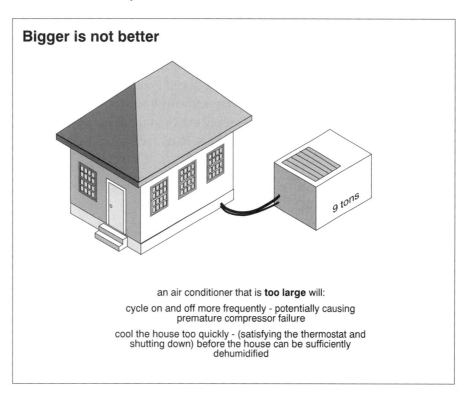

Bigger is not better

an air conditioner that is **too large** will:

cycle on and off more frequently - potentially causing premature compressor failure

cool the house too quickly - (satisfying the thermostat and shutting down) before the house can be sufficiently dehumidified

4.2 CONDITIONS

These are the common capacity issues:

1. Undersized
2. Oversized

4.2.1 UNDERSIZED

Causes
Undersized air conditioners may result from poor installation practices that do not include a heat gain calculation or do not adequately recognize the characteristics of the home. Undersized units may also be a result of house changes or additions. For example, the addition of skylights or the removal of mature trees can increase the heat gain dramatically.

SECTION ONE: AIR CONDITIONING

Implications During moderate weather, the air conditioner may function adequately, but during hot weather, the air conditioner may not be able to achieve a 15°F to 20°F temperature differential between indoors and outdoors.

Strategy The first step is to determine the size of the air conditioning system. This can often be done by reading the model number on the data plate. This is typically located on the outdoor (condenser) unit. The size may be recorded in thousands of BTUs per hour, or in the number of tons.

Carrier's Blue Book Sometimes it is difficult to translate a model number into a system capacity. The **Carrier Blue Book** available through ASHI® or Carrier Corporation in Indianapolis is an excellent reference guide with the model, serial numbers and SEER (Seasonal Energy Efficiency Ratings) of many residential air conditioning systems used in the United States.

Guessing The Size If the size cannot be determined from the model number on the data plate, the size can be approximated from the Rated Load Amperage (RLA) on the data plate. A typical reciprocating compressor will be rated at 6 amps to 8 amps per ton of cooling. The newer high-efficiency units and scroll compressors will draw less current, more like 5 amps per ton. Be sure to make it clear that this is an approximation only.

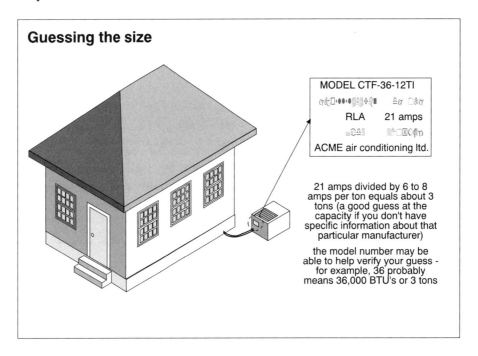

Use House Square Footage The next step is to roughly calculate the above-grade square footage of the home. Divide the square footage into the number of tons and determine the number of square feet per ton.

If the number of square feet per ton exceeds the ranges we discussed, it is probably best to describe this as marginal or suspect capacity and to recommend further investigation. There may be a number of factors in the home which cause the guidelines not to apply.

SECTION ONE: AIR CONDITIONING

Guidelines

It's also possible to find a system that seems to be just fine with respect to capacity using your guideline and yet it isn't really big enough. When considering the square footage of the house, the basement is not usually considered. However, if the basement has a walk-out with a large glass surface facing south, east or west, the air conditioning load may be far greater than contemplated.

Measure Temperature Drop Across Indoor Coil

If the system is adequately sized and is working properly, the air temperature entering the evaporator coil will be whatever the room temperature is. Let's say it's 75°F. The air coming off the coil should be 14°F to 22°F cooler (some say 15°F to 20°F). If the inlet temperature is 75°F, the air coming off should be 55°F to 60°F. This can be measured with a thermometer with a sharp probe that is pushed into a joint or hole in the supply plenum immediately downstream of (or after) the evaporator coil.

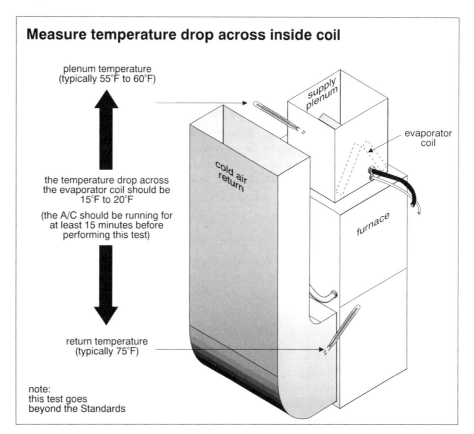

Measure temperature drop across inside coil

plenum temperature (typically 55°F to 60°F)

the temperature drop across the evaporator coil should be 15°F to 20°F

(the A/C should be running for at least 15 minutes before performing this test)

return temperature (typically 75°F)

note: this test goes beyond the Standards

If the temperature drop is different, the problem may be size-related or may indicate a need for servicing. This test should be compared with your approximation of the size of the air conditioner, based on the number of square feet per ton. Make sure the temperature drop is measured after the system has established equilibrium. The unit should run for at least 15 minutes before checking the temperature split.

Note: Measuring this temperature split is beyond the Standards but is mentioned because many inspectors do it.

4.2.2 OVERSIZED

An oversized air conditioner is susceptible to short cycling, inadequate dehumidification and large temperature variations in the house.

Causes Oversized air conditioners are usually the result of a design or installation problem.

Implication Oversized units will have a shortened life expectancy and will provide a less comfortable environment. The largest comfort issue is the lack of dehumidification. Because the temperature drops rapidly with an oversized unit, there is not an adequate volume of air movement across the coil to extract the water from the house air. This results in a house that is cold, but with a humid, swamp-like environment. Since compressors experience most damage on start-up, short cycles also mean more start-ups and a shorter life.

Strategy Other than the rough guideline test, it is difficult to know whether and how much the unit is oversized. Some public utilities indicate that a unit may be as much as 25 percent oversized without adverse effect. The temptation to oversize may become apparent when we talk about heat pumps. Since heat pumps have to deal with a much large temperature differential from outside to inside, the tendency is to make the heat pump large enough to meet the heating demand. This makes it too large for the cooling load. There are some strategies to address this problem, but within this context, we are watching for oversized cooling units.

One way inspectors identify an oversized air conditioner is by sensing the cold damp environment when walking into a house. Also, an air conditioner that short cycles (turns on and off every five minutes) is a suggestion that the unit may be oversized.

Two surveys have shown that one third to one half of all residential air conditioing systems are oversized.

Summary

While the Standards don't require it, most inspectors will red-flag systems that seem too big or too small. They will usually phrase it as a question rather than a conclusion.

Now, let's look at the individual components.

SECTION ONE: AIR CONDITIONING

▶ 5.0 COMPRESSOR

5.1 INTRODUCTION

Function
The compressor is the heart of the air conditioning system. It is the pump that drives the Freon through the system.

Location
On conventional split-system air conditioning systems, the compressor is located in the condenser cabinet outdoors or in the attic. It is typically a large, black, hermetically sealed cylinder.

The motor and compressor operate in a Freon-filled environment. We've already discussed the fact that opening the condenser cabinet to get at the compressor is going beyond the Standards.

Condensers Hung On Houses
Most condensers sit on the ground but some have wall-mounted brackets for hanging the units off the building. This keeps them out of the dirt and eliminates settling problems. Some claim that this transfers noise and vibration into the house.

Internal Protection
Compressors have internal devices to protect against such things as electric motor overload or locked rotor, improper voltage supply, high temperature or pressure, low refrigerant pressure and short cycling. These devices don't always work to protect the compressor against the many things that can go wrong.

Low Temperature Limitations
Compressors shouldn't be tested when the outdoor temperature is below 65°F or when electrical power has been on for less than 12 to 24 hours. Damage can be done to the compressor. The refrigerant in the compressor can mix with the lubricating oil in the base of the compressor. This mix does not provide good lubrication and the compressor may seize.

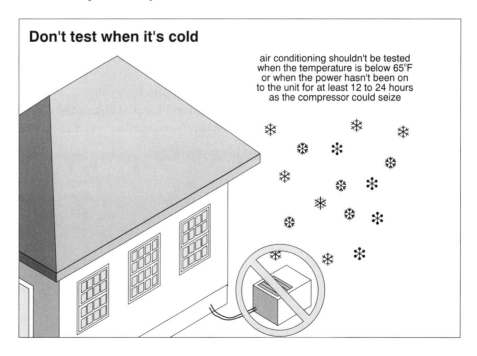

49

SECTION ONE: AIR CONDITIONING

Sump Or Crankcase Heater

There is a sump heater (crankcase heater) on many air conditioning systems that keeps the oil at the base of the compressor warm enough to boil off the refrigerant. The heater may be internal or it may be an external ring-heater wrapped around the compressor base.

Getting The Refrigerant Out Of The Oil

This heater is usually on whenever the air conditioning system is powered, but is off when the unit is shut down for the heating season. It's a waste of electricity to keep this heater on year round. It may take this heater 12 to 24 hours to warm the oil to the point where all the refrigerant is boiled off. That's why you can't just turn the power on and start up the air conditioner.

Beyond The Standards

The heater will make the base of the compressor feel warm even when the compressor is not running. Some inspectors check the temperature of the compressor base before turning on the system, to ensure that the sump heater is on. This requires removal of the condenser unit cover and is beyond the Standards.

Compressor Types

There are five types of compressors, including –
- reciprocating (piston)
- scroll
- rotary
- centrifugal, and
- screw

Residential units are typically reciprocating, rotary or scroll types. This isn't something you need to determine during the inspection.

Delayed Start-Up

Compressors that use capillary tubes as expansion devices are typically low-torque units. This means that on start-up, they expect to see equal pressure on the suction and discharge side of the compressor. The capillary tube allows this equalization of pressure but it takes a little time. When the compressor shuts off, it has a high pressure on the discharge side (as high as 275 psi from Freon 22) and a low pressure on the suction side (typically 70 psi with Freon 22).

Compressor Damage

If the compressor is asked to start again immediately, it will be trying to push against a high pressure. This can damage the compressor. As a result, most air conditioning systems have a five minute delay built in that prevents the compressor from coming on just after it has stopped.

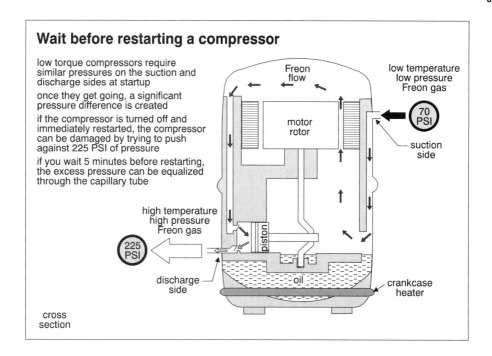

Refrigerators And Freezers Too

Incidentally, if you unplug refrigerators or freezers during inspections, you should wait five minutes before plugging them back in, for the same reason. But please, don't forget to plug them back in!

5.2 CONDITIONS

Common compressor problems include the following:

1. Excess noise/vibration
2. Short cycling or running continuously
3. Out of level
4. Excess electric current draw
5. Wrong fuse or breaker size
6. Electric wires too small
7. Missing electrical shutoff
8. Inoperative
9. Inadequate cooling

5.2.1 EXCESS NOISE/VIBRATION

Compressors are not silent when operating. They should run with a monotonous drone. Knocking sounds coming from the compressor are cause for concern. A hum with no background compressor noise may indicate that the compressor is inoperative. No noise at all also indicates that the compressor isn't working.

SECTION ONE: AIR CONDITIONING

Normal Noises
Scroll-type compressors have a different sound than reciprocating compressors. They have a higher pitched whine. It takes some experience to determine what is typical compressor noise. Some compressors are noisy when new and will break in over time.

Vibration
Some vibration is common with any compressor. Compressors are mounted on rubber feet to isolate this vibration. Excessive vibration, often accompanied by unusual noise, is a sign of severe problems with the system.

Causes
Compressor noise and/or vibration may be the result of –

- bad valves, pistons or bearings
- a poorly secured compressor (e.g., a broken compressor mount)
- slugging (trying to pump liquid Freon or oil)

Implications
Unusual noises often indicate imminent failure of the compressor.

Excess vibration leads to joint failure in the Freon line connections. This will allow Freon loss which leads to compressor burnout. Internal damage may also be done to the compressor.

Strategy
Compressor noise can be heard best if you are close to the compressor. Make sure you don't confuse the sound of the outdoor coil fan with compressor noise.

Listening With A Screwdriver
Some inspectors remove the outdoor cover and press the tip of a screwdriver against the shell of the compressor and the base of the screwdriver against the ear drum. This will transmit sounds to your ear without background noise from the fan, allowing a more accurate assessment.

Scrolls Sound Different
The word "Scroll" in the data plate will tell you that this is a scroll-type compressor. Make sure you don't misdiagnose the noise that is typical of a scroll compressor as an unusual noise.

Slugging
Compressors that are slugging may be noisy intermittently. Slugging is the introduction of liquid to the intake side of the compressor. Compressors are not intended to work on liquids. They expect to see a gas. The liquid may be oil or liquid Freon. In either case, slugging is very hard on a compressor.

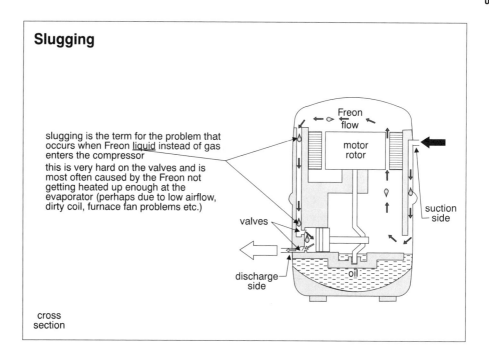

Slugging

slugging is the term for the problem that occurs when Freon <u>liquid</u> instead of gas enters the compressor

this is very hard on the valves and is most often caused by the Freon not getting heated up enough at the evaporator (perhaps due to low airflow, dirty coil, furnace fan problems etc.)

cross section

When the compressor is running, look for evidence of vibration. This can sometimes be seen through the condenser cover or through the coil. Intermittent vibration or vibration which is readily visible may indicate problems. Look at the rubber mounts to make sure they are secure.

Don't worry about vibration on start-up or shutdown. This is typical. However, a broken mount will show up clearly at start-up and shutdown.

Where noise levels or vibrations are unusual, recommend further investigation.

5.2.2 SHORT CYCLING OR RUNNING CONTINUOUSLY

Causes There are several possible causes including –

- dirty air filters, dryers or condenser coils
- too much or too little refrigerant, or contaminated refrigerant
- a restricted expansion device
- a capacitor problem
- overheating
- a defective overload protector
- inadequate oil
- valve leaks, or
- an under or oversized compressor

Implications Ineffective cooling and shortened life expectancy of compressor.

Strategy Watch for an air conditioner that never shuts off on a mild day (on a very hot day, it's likely to run continuously) or comes on and off every five minutes. Troubleshooting is beyond our scope, but recommend that the unit be serviced.

5.2.3 OUT OF LEVEL

The condensing units should be within approximately 10 degrees of level.

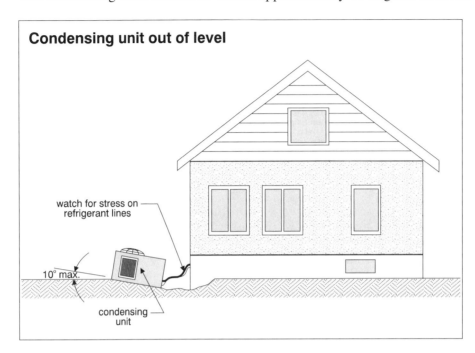

Cause Units may be out of level because of poor installation or settlement of the ground below the unit.

Implications The compressor may not be properly lubricated if the unit is not level.

The other implication of a compressor out of level is slugging. As we described earlier, this means that a slug of liquid can be suddenly released into the compressor. The compressor was not designed to compress liquids. This is very hard on the compressor.

Oil traveling with the refrigerant through the tubing may become trapped if the unit is out of level. This reduces the lubricant available to the compressor and may cause compressor failure.

Another implication of condensing units being out of level is the refrigerant lines breaking as a result of the stresses placed on lines by the unit.

Strategy Look at the condenser unit to see that it is approximately level. No tools are needed. While you are at it, check that the unit is stable and doesn't wobble. Watch for loose or rusted legs.

SECTION ONE: AIR CONDITIONING

5.2.4 EXCESS ELECTRIC CURRENT DRAW

Beyond The Standards Some home inspectors check the running amperage of the compressor condenser fan. This goes beyond the Standards, but can be done with an ampmeter. You should not do any testing of electrical devices without a good understanding of electricity. You should never put yourself at risk with respect to electric shock.

Check RLA You will want to compare the rated (some call it running) load amperage (RLA) or rated load current (RLC) from the data plate against the amperage you measure for the compressor only. In most cases, it will be roughly 80 percent of the RLA. When it's very hot outside, it might be more than 80 percent, and it could well be lower if it's cooler outside. A range of 60 to 90 percent is often considered about right.

If the running current is at or above the RLA, it is very likely that the compressor is near the end of its life. Further investigation should be recommended.

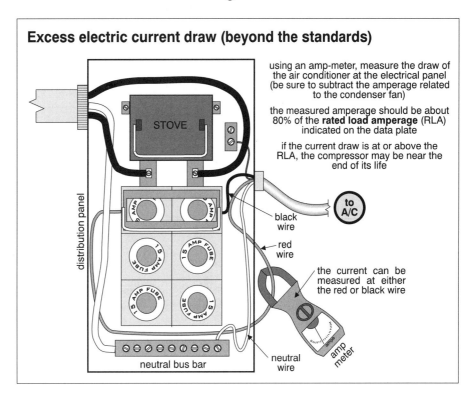

Cause The cause is compressor wear or damage.

Implication The implication is compressor failure.

Strategy When measuring the current draw, make sure that you are measuring only the compressor current. Don't inadvertently include the condenser fan current in your measurements.

SECTION ONE: AIR CONDITIONING

5.2.5 WRONG FUSE OR BREAKER SIZE

The fuse or breaker must be the right size to adequately protect the air conditioner and its wiring. The maximum fuse size is usually on the data plate. Ensure the size is correct.

Causes Inappropriate installation or servicing are the causes.

Implications Damage to the condenser unit, and possibly a fire, are the implications.

Strategy A visual check of the data plate specification against the overcurrent protection device (fuse or breaker) is all that is needed.

If the data plate gives only a fuse size rating, only fuses should be used. Circuit breakers shouldn't be used on this system. If the data plate says "HACR Breaker" (U.S. only) this means a **Heating, Air Conditioning, Refrigeration** rated breaker. Most breakers sold today are HACR rated, but look for the rating on the breaker.

No Rating Found If the data plate is missing or partly illegible, you can approximate the correct fuse or breaker size by using 125 percent of the total of the compressor RLA and the condenser fan FLA (full-load-amperage.)

5.2.6 ELECTRIC WIRES TOO SMALL

The wire carrying electricity to the condenser unit may be too small.

Causes This is an improper installation.

Implications Implications include –
• overheating of the wire and a possible fire
• excess voltage drop
• an inoperative or poor performing compressor

Strategy Check the data plate to determine the **minimum circuit ampacity**. Make sure the wires serving the condenser unit are large enough to carry this current.

No Data If the data plate is missing or illegible, you can approximate the minimum wire rating by multiplying the compressor RLA plus the condenser fan FLA (full load amperage) by 125 percent.

5.2.7 MISSING ELECTRICAL SHUTOFF

In much of the United States and since 1994 in many Canadian jurisdictions, an outdoor electrical disconnecting means is required within sight of and readily accessible to the condenser unit. Check with your local authorities to determine whether this is needed in your area.

SECTION ONE: AIR CONDITIONING

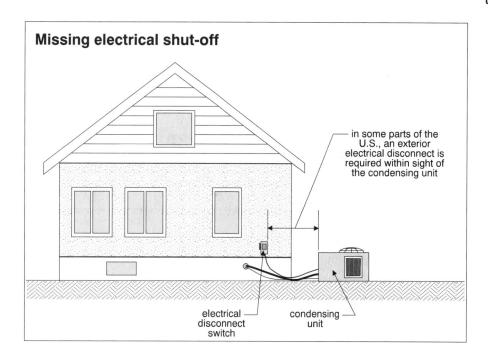

Cause	This is an installation problem.
Implication	Inability to service unit conveniently is the implication.
Strategy	Look for an accessible disconnecting means (switch) for the air conditioner close to the condenser unit. The switch should not be behind or right above the condenser unit because it may not be readily accessible.

5.2.8 INOPERATIVE

Causes Compressors may not be running because of compressor motor or pump failure, or less serious conditions including a de-energized electric circuit, starter relay failure, a weak starting capacitor, low voltage, thermostat problems or unbalanced pressure in the system.

Compressor failure may be a result of flooding, flooded starts, system contamination, overheating, a loss of oil, or slugging.

1. Flooding means liquid refrigerant gets into the oil sump (past the piston rings, for example) and messes up the ability of the oil to lubricate the bearings.
2. Flooded starts occur when liquid refrigerant accumulates in the oil sump during off cycles, often because the sump heater is inoperative.
3. System contamination may be water or other foreign materials in the refrigerant, perhaps because of careless evacuation and recharging.
4. Oil may be lost because of leakage or breakdown of the oil.

SECTION ONE: AIR CONDITIONING

5. Slugging means a slug of liquid passes through the compressor. Compressors aren't designed to receive liquids. A slug of liquid refrigerant puts tremendous pressure on the compressor valves.
6. Overheating may result from a number of causes.

Implication Obviously, an inoperative compressor means no air conditioning.

Strategy If the compressor doesn't operate when you turn the thermostat down, check first to make sure the fuse or breaker is active. Make sure the thermostat is set to the "Cool" position. If the thermostat is a mercury bulb type, make sure the switch activates by removing the cover and turning the thermostat up and down. You should see the mercury make and break contact as it moves in the bulb.

Delayed Start-Up Wait for up to seven minutes before determining that the unit is not operative. Many units have a timed start-up delay of from two to seven minutes.

It is beyond our scope to troubleshoot compressor failures. This information is intended to give you an appreciation for the complexity of the issue. If the compressor doesn't operate, just record it and recommend further investigation.

5.2.9 INADEQUATE COOLING

Cause The air conditioner may run but may not cool the house well. While the compressor may be the problem, there are several possible causes including:

1. Compressor problems
2. Condenser problems
3. Evaporator problems
4. Freon line problems
5. Expansion device problems
6. Evaporator or condenser fan problems
7. Thermostat problems
8. Airflow problems

Implications Poor comfort and high energy costs are the implications of inadequate cooling.

Strategy Measure the temperature drop across the evaporation coil, looking for a 15°F to 20°F drop, roughly. Also, sense how comfortable the house is, if inspecting on a warm day.

Summary

The compressor is the big component in air conditioners and heat pumps. Compressor problems can be difficult to diagnose. When air conditioners don't perform well, allow for the possibility of compressor problems.

SECTION ONE: AIR CONDITIONING

Air Conditioning & Heat Pumps
MODULE

QUICK QUIZ 2

☑ INSTRUCTIONS

- You should finish Study Session 2 before doing this Quiz.
- Write your answers in the spaces provided.
- Check your answers against ours at the end of this Section.
- If you have trouble with the Quiz, re-read the Study Session and try the Quiz again.
- If you did well, it's time for Study Session 3.

1. One ton of air conditioning is equivalent to how many BTUs per hour?

   _____

2. List ten items that affect the amount of air conditioning needed in a home.

   _____
   _____
   _____
   _____
   _____
   _____
   _____
   _____
   _____
   _____

SECTION ONE: AIR CONDITIONING

3. How many square feet can one ton of air conditioning cool in Florida?

   _____

4. How many square feet can one ton of air conditioning cool in Michigan?

   _____

5. Oversized distribution ductwork is a common problem with central air conditioning.
   True ☐   False ☐

6. An undersized air conditioner is better than an oversized one.
   True ☐   False ☐

7. The typical temperature drop from outdoors to indoors with a properly operating air conditioning system would be about 15°F.
   True ☐   False ☐

8. What kind of temperature drop would you expect to find in the house air as it passes over the evaporator coil?

   _____

9. A compressor can be thought of as a pump.
   True ☐   False ☐

10. Compressors are typically located indoors on an air-to-air split system air conditioning system.
    True ☐   False ☐

11. What is the function of a sump or crankcase heater?

    _____

SECTION ONE: AIR CONDITIONING

12. List nine common compressor problems.

_____
_____
_____
_____
_____
_____
_____
_____
_____

If you didn't have any trouble with this Quiz, then you are ready for Study Session 3.

Key Words:
- *One ton*
- *Ductwork capacity*
- *Undersized*
- *Oversized*
- *Temperature drop*
- *Compressor*
- *Crankcase heater*
- *Noise/vibration*
- *Short cycling*
- *Out of level*
- *Excess current*
- *Wrong breakers*
- *Undersized wire*
- *No electrical disconnect*
- *Inoperative*
- *Inadequate cooling*
- *Slugging*
- *Rated Load Amperage (RLA)*
- *Full Load Amperage (FLA)*

SECTION ONE: AIR CONDITIONING

Air Conditioning & Heat Pumps
MODULE

STUDY SESSION 3

1. You should have finished Study Session 2 and Quick Quiz 2 before starting this Session.

2. This Study Session deals with condenser coils, water-cooled condenser coils, evaporator coils, condensate systems and refrigerant lines.

3. At the end of this Session, you should be able to –

    - explain in two sentences the function of each of the following:
        - Condenser coil
        - Water-cooled condenser coil
        - Evaporator coil
        - Condensate system
        - Refrigerant line
    - list four common condenser coil problems
    - list three common water-cooled condenser problems
    - list five common evaporator coil problems
    - list one common condensate problem
    - list five common refrigerant line problems

4. This Session should take you roughly 1½ hours to complete.

5. Quick Quiz 3 is included at the end of this Session. Answers may be written in your book.

SECTION ONE: AIR CONDITIONING

Key Words:

- *Condenser coil*
- *Receiver*
- *Clothes dryer or water heater vent*
- *Water cooled condenser coil*
- *Backflow preventer*
- *A-coil*
- *Expansion device*
- *Capillary tube*
- *Thermostatic expansion valve*
- *Condensate drain pan*
- *Trap*
- *Condensate pump*
- *Suction line*
- *Liquid line*
- *Freon*
- *Filter/dryer*
- *Sight glass*
- *Sludge*

SECTION ONE: AIR CONDITIONING

▶ 6.0 CONDENSER COIL (OUTDOOR COIL)

6.1 INTRODUCTION

Function   The outdoor condenser coil transfers heat from the refrigerant into the outdoor air (or into water if it's a water-cooled system). The refrigerant enters the condenser coil from the compressor as a hot (e.g.,150°F) gas. The gas condenses to a liquid as the 95°F outdoor air (for example) is pushed past the coil by the condenser fan. The air coming off the condenser coil may be 110°F.

Beyond The Standards   The condenser coil is in a cabinet often call the **condenser unit** although it contains the compressor and the condenser fan as well as the condenser coil. This cabinet is not normally opened or serviced by the homeowner. Live 240-volt electrical connections are accessible inside the cabinet. Opening this cabinet is beyond the scope of a home inspection; however, a good deal of information can be gained by opening the cabinet. Again, this should only be done by someone familiar with air conditioner systems and experienced in working around live electrical equipment. As discussed, in some jurisdictions an outdoor electrical disconnect is required so the outdoor unit can be serviced safely.

We have talked about inspecting the compressor already. Removing the access panel allows you to examine the compressor shell and determine the compressor age from its data plate. You can also look for a failed capacitor (bulging or leaking), oil leaks (which indicate refrigerant leaks) or corroded and obstructed coil fins, for example.

Receivers   Condensers are typically copper tubes with aluminum fins. The bottom of some condensers are liquid receivers that collect the condensed refrigerant. Some receivers are separate from the condenser. Some modern coils are all aluminum; these may have a shorter life span than copper tube coils.

Noisy   The condenser unit can be noisy. Its location should take this into consideration. A condenser unit that is adjacent to a patio for example, may be a nuisance.

6.2 CONDITIONS

Common coil problems include the following:

1. Dirty
2. Damaged or leaking
3. Corrosion
4. Clothes dryer or water heater exhaust too close to condenser

SECTION ONE: AIR CONDITIONING

6.2.1 DIRTY

The outdoor coil sees unfiltered air pass through it when the system is working. Many modern systems draw air in through the sides and discharge through the top. Other intake and discharge arrangements are possible.

Causes   It's very common for some or all of the coil to be matted over with dirt and debris. The fan pulls relatively dirty air through the coil and dirt will accumulate on the upstream side. This is typically visible looking at the coil from outside the cabinet. Grass clippings discharged by lawn mowers can clog a condenser coil quickly. Leaves and wind-blown debris can also get stuck on the coil.

Implications   Restriction of air flow through the coil leads to poor heat transfer between the coil and the outdoor air. This means the refrigerant leaving the condenser coil will be warmer than it should be. This means that less heat can be removed from the system. Comfort will suffer and energy costs will be high. Over the long term, compressor damage may result.

Strategy   The condenser coil should be inspected at least from the outside of the condenser cabinet. Many inspectors use a light to ensure that there is free air passage through the coils. Where there is dirt on the coils, cleaning should be recommended as regular maintenance.

6.2.2 DAMAGED/LEAKING

It is common to find the fins on condenser coils damaged. Grilles or louvers are often damaged, and occasionally you'll find mechanical damage to the cabinet itself.

Causes   The two most common causes are lawn mowers and installers or service people. High-pressure water used to clean the coil may also damage the fins.

Implication   Damaged fins reduce airflow and heat transfer, adversely affecting system performance, comfort and costs.

Strategy   Look at as much of the fin surface as you can. If the coil is clogged, remove some of the dirt to determine if the fins are bent.

Straightening the fins is possible, but leave it to a service person.

Oil Stains   Coils may leak if they are damaged. You won't see Freon leaking (it's a colorless gas) but you may see oil stains below the condenser coil. This oil is carried with the refrigerant, and does appear on and below coils or pipes where there are leaks.

SECTION ONE: AIR CONDITIONING

6.2.3 CORROSION

Corroded fins are common, particularly on older units.

Causes Common causes include –

- excessive moisture (from soil contact, for example) or condensation (perhaps as a result of moisture trapped by non-breathing covers)
- chemicals used on lawns and gardens
- exposure to the atmosphere over time
- contaminants in the refrigerant lines

Implications The implications of a corroded coil are the same as those of a dirty or damaged coil.

Strategy The inspection strategy is the same as it is for looking for a dirty coil or damaged fins. Again, some inspectors use a light to look through the coil.

Make sure the condenser is sitting on a pad so that it is above the soil. A cabinet that is partially buried in dirt will be subject to accelerated corrosion.

Look for corrosion on the cabinet and its grilles or louvers.

6.2.4 CLOTHES DRYER OR WATER HEATER EXHAUST TOO CLOSE TO CONDENSER

It is common to find a condenser clogged with lint from a clothes dryer discharging nearby. The high temperature air from a dryer or a sidewall-vented gas-fired water heater can reduce cooling efficiency. The water heater exhaust may also corrode the coil.

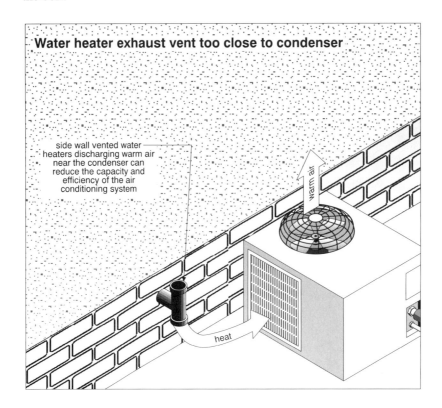

SECTION ONE: AIR CONDITIONING

Cause The condenser fan pulls the dryer or water heater exhaust into the condenser coil.

Implications The lint will clog the fins and the warm air from the dryer or water heater makes it hard to dump the heat outdoors. The fins may be corroded by the acidic water heater exhaust.

Strategy Dryer and water heater vents should be about 6 feet from the condenser. The exact distance varies with condenser configurations, elevation differences, etc. Look for lint on the coil. Recommend moving the vent if it's too close to the condenser.

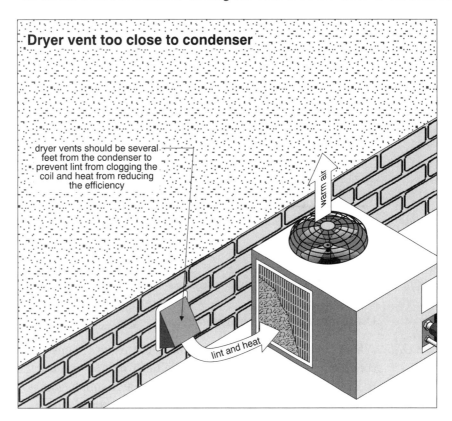

▶ 7.0 WATER-COOLED CONDENSER COIL

7.1 INTRODUCTION

Function In a water-cooled air conditioner, the heat in the Freon that has been collected from the house is not discharged into the air outside, but is discharged into water. This water may be from the city supply, from a river, lake, stream or well. In a commercial application, the water might be continuously cycled through a cooling tower. In single-family residential applications, the water just passes by the condenser coil once and is discarded.

Indoor Condenser The condenser coil is usually located near the evaporator coil, since it does not have to be outdoors or in the attic to discharge heat into the outside air. The condenser sees a more predictable, uniform and dryer climate, and some say this leads to a longer life expectancy. On the downside, the heat given off by the compressor motor ends up inside the house. Obviously this doesn't help with air conditioning.

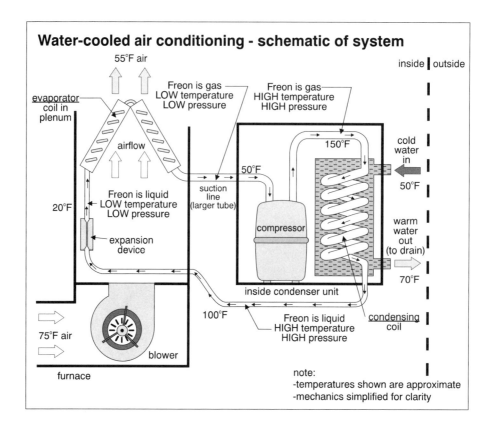

Double Pipe Condenser

Water-cooled coils do not employ fins, since we are not passing air over the coil. It is typically a jacketed system with the refrigerant lines inside the water tubing. Usually the water and refrigerant flow in opposite directions.

Uses Lots Of Water

Where city water is used, many consider this an expensive and inefficient air conditioning system. Some municipalities won't let you use this kind of system because it wastes so much water. The water often simply goes down the drain after it has picked up heat from the condenser coil. In some applications, the water can be diverted to be used in watering the lawn or filling a swimming pool. The supply water to the coil should never be from the pool, however. The chlorine in the pool water will corrode the coil quickly.

7.2 CONDITIONS

Common water-cooled coil problems include the following:

1. Leakage
2. Coil cooled by pool water
3. No backflow preventer
4. Low plumbing water pressure

SECTION ONE: AIR CONDITIONING

7.2.1 LEAKAGE

Causes
- Leaks may occur in the water-cooled coils because of chemical reactions between the cooling water and the jacket material.

- Leaks may also be the result of mechanical damage, including vibration. In some cases, the coils are enclosed in a metal cabinet, but in others the coils are exposed.

- Corrosion as a result of chemicals in the home may also cause leakage. Paint stripping, for example, can create a very corrosive environment. Swimming pool chemicals in poorly sealed containers can have a similar effect.

Implications Non-performance of the air conditioning is one implication, and if the leakage goes undetected, extensive water damage can result.

Strategy Look for evidence of leakage, including stains, moisture trails or wetness. If the leak is downstream of a solenoid valve that is activated when the air conditioner is working, there may be no moisture apparent when the unit is at rest.

Before starting the unit, ensure that any supply water isolating valve is open. Water must be flowing for the system to run.

A system operating properly will have discharge water that is 15°F to 20°F warmer than the inlet water.

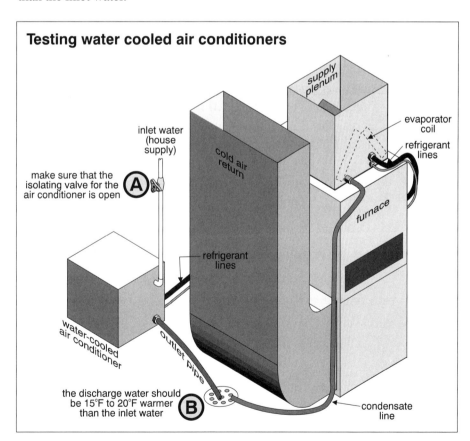

Testing water cooled air conditioners

SECTION ONE: AIR CONDITIONING

7.2.2 COIL COOLED BY POOL WATER

Some installations circulate water from the swimming pool through the cooling jacket and back to the swimming pool. This strategy uses heat from the house to help warm the swimming pool water. The consensus is that this is a bad idea because the chemicals common in swimming pool water may attack the cooling jacket and cause premature failure.

Causes — The decision to recirculate pool water through this jacket is a design and installation issue.

Implications — Corrosion, leakage and premature replacement of the cooling coil are the implications.

Strategy — Trace the supply and discharge water lines from the jacket, if possible, to determine where the source water is from. If it is from the pool or you suspect it is from the pool, recommend further investigation by a specialist and let the client know that an alternate arrangement may be recommended.

Don't be fooled by water discharging to the pool. Bringing fresh water from the municipal supply, for example, and discharging the heated water to the swimming pool does not expose the jacket to water that contains pool chemicals.

7.2.3 NO BACKFLOW PREVENTER (ANTI-SIPHON DEVICE)

Where the cooling jacket is fed by the house water supply, an anti-siphon device such as a backflow preventer may be required by the local plumbing authority so that water cannot go back into the drinking supply after passing through the cooling jacket. Where this backflow preventer is not in place, a **cross connection** exists. This is a dangerous situation in which the drinking water may be contaminated. Check in your area to see whether or not a backflow preventer is required.

SECTION ONE: AIR CONDITIONING

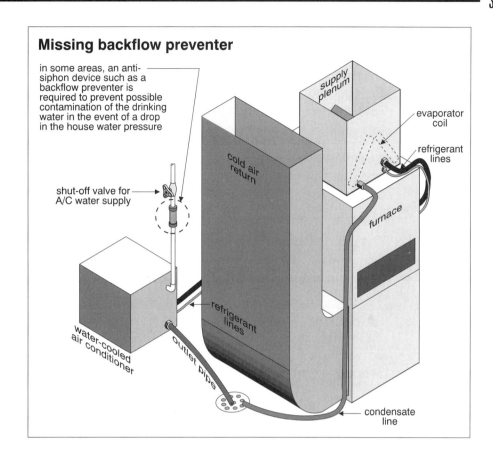

Missing backflow preventer

in some areas, an anti-siphon device such as a backflow preventer is required to prevent possible contamination of the drinking water in the event of a drop in the house water pressure

Causes	The omission of a backflow preventer may be an installation mistake.
Implications	Possible contamination of potable (drinking) water is the implication.
Strategy	Look for a backflow preventer on the supply inlet to the air conditioner, when the supply is from the house drinking water.

7.2.4 LOW PLUMBING WATER PRESSURE

Cause	Water-cooled units can use so much water, they may reduce house water pressure drastically. Water-cooled air conditioners typically need about 3 gallons per minute (gpm) flow for every ton. A three-ton unit may need 9 gpm! This doesn't leave much for taking showers.
Implication	The water pressure for the house plumbing system may not be adequate when the air conditioning is running.
Strategy	If possible, do your plumbing system flow tests with the air conditioner running.

► 8.0 EVAPORATOR COIL (INDOOR COIL)

8.1 INTRODUCTION

Function The evaporator coil transfers heat from the house air into the refrigerant in the coil. The refrigerant boils inside the coil and the house air temperature drops (e.g., from 75°F to 60°F) as it is pushed past the coil by the house air fan.

Copper Tubes Evaporator coils are usually copper tubes with up to 14 aluminum fins per inch of tube. This type of coil provides good surface area for heat transfer and allows the coils to be compact. Some modern systems have all aluminum coils, which may have a shorter life than copper tube coils.

A-Coils This coil is often an **A-coil** installed with a condensate tray below the bottom of each leg of the "A". These trays catch the condensate coming off the coil and allow air movement through the coil. The air passes up between the trays through the coil.

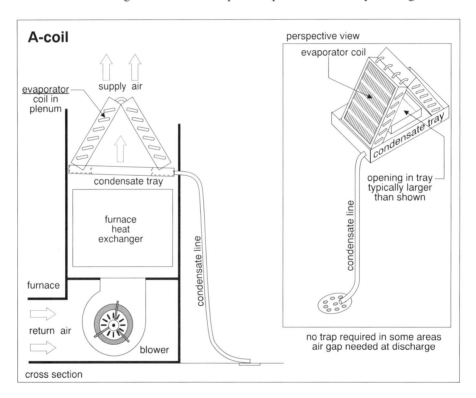

Slab Coils Slab coils are installed on a small angle with a single condensate tray below the lower end of the coil.

SECTION ONE: AIR CONDITIONING

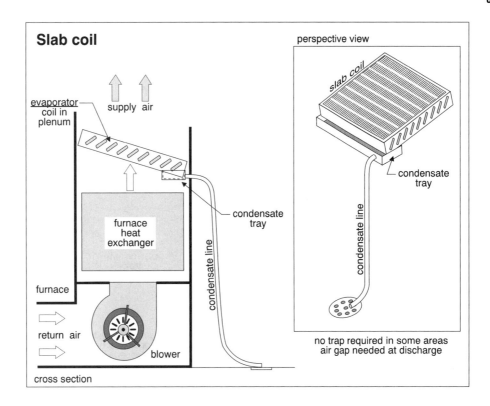

Vertical Coils If the unit is arranged so that airflow across the coil is horizontal, the coil can be vertical with a tray at the bottom.

Coil Location Houses with forced-air heating typically have the evaporator coil just past (downstream of) the furnace heat exchanger. The coil may be in a basement, crawlspace, attic or closet.

Houses without a forced-air furnace are often also houses without basements. The evaporator coil, fan and much of the duct system are often in the attic.

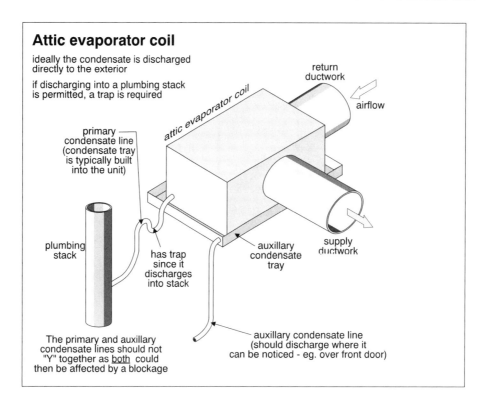

Downstream Of Furnace Heat Exchangers

Evaporator coils must be downstream of (after) gas, oil or propane furnace heat exchangers. Heat exchangers would rust out quickly if the coils were upstream. The coils can be either before or after electric heating elements.

SECTION ONE: AIR CONDITIONING

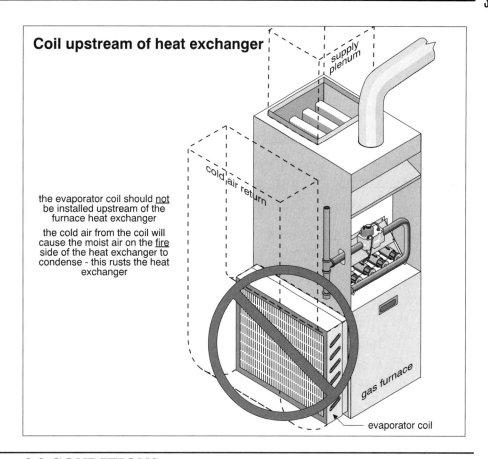

8.2 CONDITIONS

Common evaporator coil problems include the following:

1. No access to coil
2. Dirty
3. Frost
4. Top of evaporator dry
5. Corrosion
6. Damage

8.2.1 NO ACCESS TO COIL

On many installations, there is no access port in the ductwork to inspect and service the coil.

Causes This is an installation issue.

Implications If the unit can't be accessed for inspection and cleaning, it will get dirty and suffer from a lack of maintenance. It's also possible that condensate overflow or leakage may occur and go undetected for some time. If the coil is above the furnace heat exchanger (common in an upflow furnace), considerable rusting of the heat exchanger can occur.

Strategy In some areas, the coils are always accessible. In other areas, coils are rarely installed with an access cover. Find out what is the norm in your area, and if the coils are usually accessible, write up any that are not. Whether the coils are typically accessible or not, it's probably safe to say they need cleaning if there is no access.

When coils are in the attic, access is often awkward and unpleasant, especially on a hot day. However, it is part of a Standard inspection to check these.

8.2.2 DIRTY

Evaporator coils have fins which are easily clogged.

Causes Dirt may accumulate on the fins as a result of a missing or dirty filter. Even if the filter is in place and is changed regularly, dirt will eventually accumulate on the coil. Filters in attics are often not changed regularly.

Implication Dirt on the coil restricts airflow through the coil and inhibits heat transfer across the coil, resulting in inadequate cooling and high operating costs.

Another implication is the possibility that the Freon returning to the compressor may be too cold and in a liquid state. The compressor may be damaged if it pumps liquid rather than gas.

Strategy Look at the coil to see if it's clean. If possible, look at the upstream side of the coil where the dirt will typically accumulate. Sometimes a mirror and flashlight are necessary to get a good look. The dirt often accumulates in a mat which can be rolled off the coil, much like the lint that accumulates on a clothes dryer filter.

In some cases, mold and mildew may also be noted on the coil.

SECTION ONE: AIR CONDITIONING

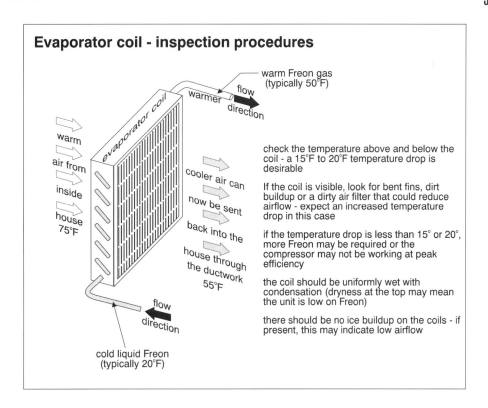

8.2.3 FROST

If frost is seen on the coil, service is necessary.

Causes An improper refrigerant charge may result in frost on a coil. Inadequate air flow across the coil also results in a frost build up. This reduces the ability of the coil to transfer heat. Eventually, the coil will frost over completely. There are several possible causes of inadequate air flow, including –

- inoperative fan
- dirty fan blades
- dirty filter or evaporator coil
- fan belt slipping
- undersized or obstructed ducts
- bypass humidifier with an open damper (this is further discussed in the Duct System section later on)

Implication System performance will deteriorate and compressor damage may result from the suction line containing liquid rather than gas.

Strategy Check the coil for frost. Although no frost should be visible, if there is a small amount of frost on the inlet port of the coil, this may not be a serious problem. Servicing should, however, be recommended.

If frost is visible on other parts of the coil, compressor damage is a distinct possibility and the system should be shut down and recommended for service.

8.2.4 TOP OF EVAPORATOR DRY

When the evaporator is operating, the entire surface should be covered with condensation.

Cause If the top of the coil is dry, this may indicate that the unit is starved for refrigerant.

Implications Poor comfort and system stress are the implications.

Strategy If you can look at the coil when the system is or has been operating, check for uniformity of condensation on the coil. This is a secondary indicator to checking the temperature split across the coil. If the split is appropriate, there may not be a problem.

8.2.5 CORROSION

The copper or aluminum fins may corrode.

Causes Causes of corrosion on the coil include a chemical reaction between the refrigerant and the refrigerant lines, contaminants in the system, acidic condensate, blocked condensate drain, etc. Corrosion can also result from household activities, such as paint stripping or open storage of swimming pool chemicals. You won't be responsible for determining the cause of the corrosion, but you should be able to identify it.

Implications All coils will corrode as they age, and as long as there is not a significant reduction of heat transfer or airflow, this is not a major concern.

Corrosion caused by system contaminants is much more serious. These contaminants are usually left over from an improper evacuation of the refrigerant when a component was replaced. The contamination leads to refrigerant line failure, expansion device blockage and corrosion of the soldered joints.

Strategy The coil should be examined through the access port, paying particular attention to the joints at the expansion device and the connection between the refrigerant tubing and the coil. Corrosion of the copper will be green or blue. If the problem is system contamination, the corrosion will be uniform around the joint.

8.2.6 DAMAGE

The fine aluminum fins on the coils can be damaged by aggressive cleaning. If the fins are bent, the airflow will be restricted.

Cause Coils are often mechanically damaged during cleaning.

Implications Reduced air flow and comfort in the house, reduced heat transfer and possible compressor damage are the implications.

Strategy When inspecting the coil, look for damage to the fins. In some cases, damaged fins can be straightened.

SECTION ONE: AIR CONDITIONING

8.3 EXPANSION DEVICE (METERING DEVICE)

Function This is the device located just upstream of the evaporator coil that changes the refrigerant from a high pressure, high temperature liquid to a low pressure, low temperature liquid.

Capillary Tube The most common expansion device in residential air conditioners is a **capillary tube.** This is a very small-diameter copper tube wrapped into a coil. It simply creates a bottleneck in the liquid line. As the refrigerant comes out of the end of the small-diameter capillary tube, it expands into the larger tube. This lowers the pressure and, consequently, the temperature.

Thermostatic Expansion Valve Another type of expansion device is the **thermostatic expansion valve (TXV or TEV).** The TXV is usually found on air conditioning systems larger than three tons, or on combination heat pump and electric furnace packages. The TXV is a more precise metering device than the capillary tube and allows the system to adjust to its environment so that the compressor should never receive liquid Freon.

Function The TXV controls the refrigerant flow into the evaporator coil by sensing the heat from the refrigerant flowing out of the coil. A sensing bulb is mounted on the downstream side of (after) the evaporator coil, and is connected to the TXV by a fine copper tube.

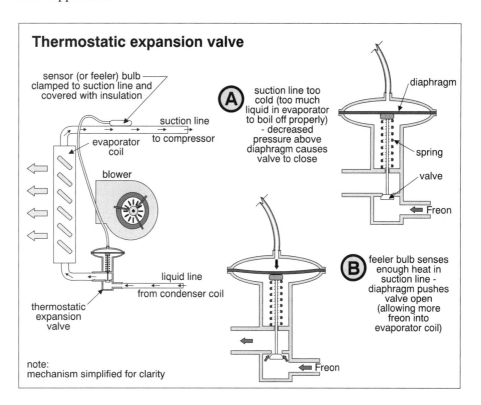

Location The expansion valve is often visible in the refrigerant lines just outside the duct system, close to the evaporator coil.

SECTION ONE: AIR CONDITIONING

8.3.1 CONDITIONS

Common expansion device problems include the following:

1. Capillary tube defects
2. Thermostatic expansion valve connections loose
3. Clogged orifice
4. Expansion valve sticking

8.3.1.1 Capillary tube defects

The capillary tube may be crimped, disconnected, frosted, or covered with oil.

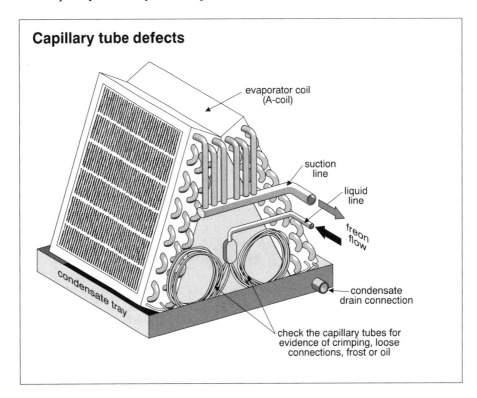

Causes   These conditions may indicate mechanical damage, damage caused by vibration, Freon leaks, or a restriction in the tube.

Implication Poor cooling performance will result.

Strategy   If visible through the access port in the duct system, look at the tube for evidence of crimping, loose connections, frost or oil.

8.3.1.2 Thermostatic expansion valve connections loose

The TXV may have a loose bulb or loose liquid line connection. The liquid line connection point may also be cracked.

Causes Vibration, under-tightening or over-tightening on original installation may be the cause.

Implication Loss of Freon and poor system performance will be the implications.

Strategy Look for evidence of a loose bulb or liquid line connection. Look for evidence of cracking at the liquid line connection.

8.3.1.3 Clogged orifice

Cause The capillary tube or TXV can become obstructed by foreign objects, ice, wax or sludge in the refrigerant. Sludge develops as a result of oil or refrigerant breakdown.

Implications If the expansion device is plugged, the refrigerant can't flow and the system will shut down, ideally before the compressor is damaged.

Strategy Troubleshooting air conditioning systems is beyond our scope, but frost near the expansion device may suggest an obstruction.

8.3.1.4 Thermostatic expansion valve sticking

The valve components may not move properly.

Causes This is caused by valve malfunction or sludge.

Implications No refrigerant flow is the implication.

Strategy Again, this may not be identified as the reason for poor system performance or non-performance, but this is beyond our scope.

SECTION ONE: AIR CONDITIONING

▶ 9.0 CONDENSATE SYSTEM

9.1 CONDENSATE DRAIN PAN (TRAY)

9.1.1 INTRODUCTION

Location And Function

The **condensate drain pan**, located below the evaporator coil, catches the condensate that drips off the outside of the coil as the warm moist house air passes over the coil. The cold refrigerant in the coil cools the air, causing the moisture to condense on the outside of the coil. This is how air conditioners dehumidify the house air.

Pan Gathers Condensate

The condensate that collects must be carried away. A three-ton air conditioner can generate two gallons of condensate per hour on a humid day. The first step is the gathering of the condensate as it drips off the coil. If there is access to the coil, this pan can be inspected. If there is no access, you won't be able to see the pan, and as mentioned earlier, you should recommend servicing and the provision of an access panel so that the coil and pan can be inspected and cleaned. The pan should be slightly sloped so water flows to the drain connection.

The condensate is piped to a floor drain, sink, outdoors or another suitable discharge point.

9.1.2 CONDITIONS

The common drain pan problem is –

• leaking

9.1.2.1 Leaking

The pan may be leaking or overflowing for several reasons:

Causes
- The pan may be cracked or have an open seam that allows water to run out.
- The pan may be substantially filled with dirt or debris that limits the amount of water it can carry.
- The opening to the drain line may be obstructed.
- The drain line may be missing or disconnected.
- The slope of the pan may be inappropriate, allowing water to overflow rather than run to the drain.
- The pan may be rusted and holes may have developed.
- The pan may not be well secured in place or may not be directly under the low point of the coils.

SECTION ONE: AIR CONDITIONING

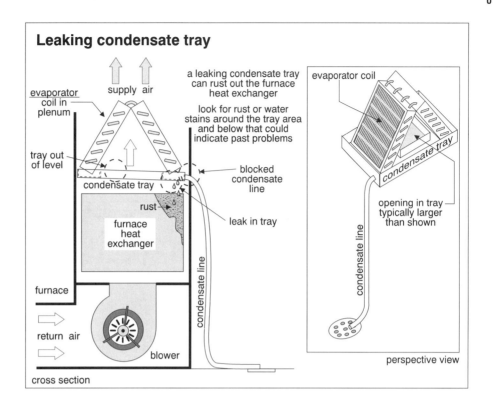

Leaking condensate tray

Implications	Leaking pans will allow water damage to occur to whatever is below. In many areas, the evaporator coil is directly above the furnace heat exchanger. A leaking pan can rust out the furnace heat exchanger, rendering the furnace ineffective and creating an unsafe situation, with a risk of carbon monoxide entering the house.
Strategy	If there is an access panel, look at the pan for evidence of leaking, including cracks, rust, dirt accumulation, a drain line obstruction, improper slope or poor attachment. Look for staining on the outside of the pan and on whatever components are below.
	If there is no access cover, look around and below the cabinetry and furnace components below the coil for evidence of water streaking, staining or rusting.
	If the air conditioner is running or has been running recently, look for moisture.

9.2 AUXILIARY CONDENSATE DRAIN PAN

Second Line Of Defense   Auxiliary pans are typically provided where an evaporator coil is located in an attic or anywhere above finished living space. These are only used when failure of the main drain pan will damage the house. This is a secondary pan, provided because the primary drain will eventually leak.

Causes Of Leaks   This pan should be checked for the same conditions as were discussed in 9.1, including rust, dirt accumulation, a blocked drain line, improper slope and poor attachment.

Separate Drain Pipe Needed

The auxiliary drain should have a separate drain line from the pan that is not manifolded with the line from the primary pan. A clog in one condensate drain line should not render both pans ineffective. There should be no trap in the auxiliary drain line, which is typically ³/₄-inch diameter. The drain should terminate in a spot where discharge will be noticed, since it is an emergency overflow. Over the front door is good.

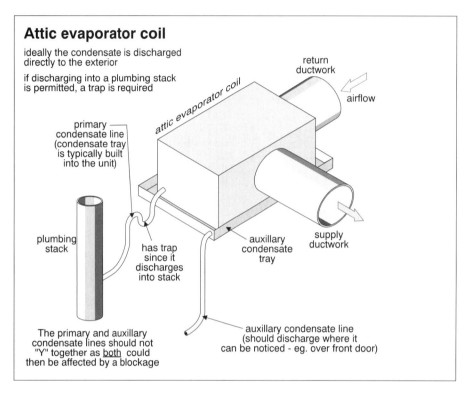

Float Switch

Some drain pans have a float switch that will shut off the system if the water level rises too high, instead of an auxiliary pan. This switch may shut off just the compressor or the compressor and the blower. This switch can be tested by holding it up for roughly five minutes to see whether the unit shuts off. In some cases the whole system shuts down. In some cases only the compressor and condenser fan shut down.

9.3 CONDENSATE DRAIN LINE
9.3.1 INTRODUCTION

Function And Material

This drain line carries condensate from the pan to an appropriate discharge point. The line is typically plastic or copper and is usually ³/₄-inch diameter. In some areas, the condensate line is insulated to prevent it from sweating as it runs from the evaporator coil to the discharge point.

Traps

The line from a primary pan is sometimes trapped. The line from a secondary pan should not be trapped.

Discharge Point

The discharge point from a primary pan should be out of the way. The discharge point for a secondary pan should be in the way, and hard to ignore. It should catch your attention, because something is wrong if it's working.

SECTION ONE: AIR CONDITIONING

9.3.2 CONDITIONS

Common drain line problems include these:

1. Leaking
2. Blocked
3. No trap
4. Improper discharge point

9.3.2.1 Leaking

Causes   The condensate line may be –

- disconnected from the condensate tray
- split
- disconnected at the trap
- crimped, clogged or
- mechanically damaged anywhere along its length.

Implications   A leak will allow condensate to flow where it was not intended. The implications are similar to those we discussed with respect to leakage from the condensate drain pan.

Strategy   Inspect the length of the condensate drain line looking for active leakage if the unit is running or has been running recently. If the air conditioner is idle, look for evidence of staining, streaking, loose connections or split piping.

Beyond The Standards   If there is access to the drain pan, some inspectors pour water into the pan and watch the discharge point to ensure that water comes out here. Leakage should show up along the length of the pipe. This goes beyond what most home inspectors would do.

9.3.2.2 Blocked

Cause   The drain line is subject to blockage by insects and debris.

Implications   The implications are similar to those caused by a leak in the drain pan.

Strategy   If the unit is operating, make sure water is discharging out of the drain line. If the unit is at rest, pouring water into the drain pan and checking the discharge point will reveal blockages, although this goes beyond the Standards. In some cases, there is a vent (relief opening) downstream of the trap in the condensate drain line. Watch that this is not discharging. If water is coming out of this relief opening, there is probably a blockage in the trap.

Blockages are common at the entry and exit points at the drain line.

9.3.2.3 No trap

Traps are often required on condensate drains, especially when the condensate pan is upstream of (before) the house air fan. This is typical of air-conditioning-only systems, or air-conditioning-plus-electric-heat. The water in the pan is under negative pressure because it is on the suction side of the pan.

Purpose — Negative Pressure

If the line is under negative pressure, air may be sucked up the drain and the water may not be able to flow down the drain line. In some cases, the drain pan may fill up and overflow. A trap prevents this. Standard traps have a 4-inch drop on the upstream side, and a 2-inch rise on the downstream side.

Sometimes Not Required

Traps in condensate lines are required in many jurisdictions. In some jurisdictions, if the condensate line is permitted to drain into a waste plumbing pipe, a trap will be required. Check with authorities in your area to determine whether air conditioning condensate lines require traps and, if so, whether there are any specific rules about where and how they are installed. Traps are usually outside the evaporator plenum, as close to the drain pan as possible.

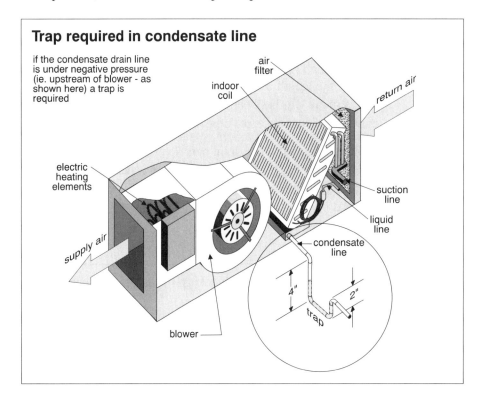

Not Into Plumbing Stack

If the condensate line does terminate in a waste plumbing stack, the trap will prevent sewer odor and bacteria from entering the house air supply, at least until the trap water evaporates. This discharge point is not allowed in most areas.

SECTION ONE: AIR CONDITIONING

Causes — Missing traps have usually been omitted during installation.

Implications — Some systems without traps work most of the time. During hot humid weather, the drain pan may overflow. Most would agree that if a condensate line discharges directly into a waste plumbing system, a trap is necessary for health reasons. Incidentally, when the air conditioner hasn't been run for several months, the water in the trap will evaporate, and for this reason, direct connection to a waste plumbing stack is not a good idea even if permitted in your jurisdiction.

Strategy — Trace the condensate line from the drain pan to its discharge point. It should be obvious whether or not a trap has been provided.

No Water In Trap — When the air conditioning system is idle, the water in the trap will evaporate, rendering the trap ineffective. This is not a defect.

9.3.2.4 Improper discharge point

Discharge Outside — The best place for air conditioning discharge is usually considered to be outside the building onto the ground. The discharge should not be outside the building part way up the wall so the condensate runs down the exterior of the wall.

Sink, Floor Drain, Plumbing Fixture — Some jurisdictions allow the condensate to drain into a sink or floor drain as long as there is an air gap to prevent siphoning. It may be permitted to discharge the condensate into the tail piece of a basin (upstream of the trap) or into the overflow for a bathtub (again, upstream of the trap). Check with authorities to find what is allowed in your area.

Through Floor — In some areas, the condensate drain line goes down through the floor slab and discharges into the granular material below. This is not considered a good arrangement because you can't see the condensate line working. An obstruction may go undetected until considerable damage is done. Sub-slab gases (such as radon) may also find their way into house when the unit is not running.

Into Waste Plumbing Stack — We also recommend against discharging the condensate drain line into the waste plumbing stack. There is the issue of a blockage going unnoticed for some time and there is also the possibility of sewer odor backing up into the house air system, as discussed earlier.

SECTION ONE: AIR CONDITIONING

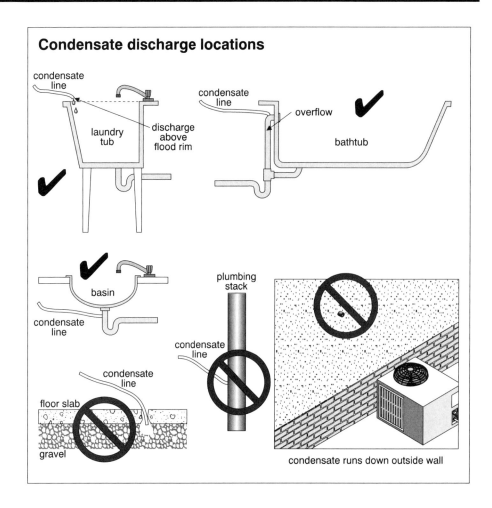

Causes Improper condensate line discharge is an installation issue.

Implications Non-performance that goes unnoticed may cause water damage. If discharging into waste plumbing, sewer odor and its associated health issues are other potential implications.

Strategy Trace the condensate discharge line to its termination point. Ensure that the end of the line is open, and if the air conditioner is operating, water should be coming out.

9.4 CONDENSATE PUMP

9.4.1 INTRODUCTION

If the condensate cannot run from the evaporator coil outside by gravity, and if there is no drain or sink nearby that the condensate can be discharged into, a pump is required. Condensate pumps are typically low-quality sponge pumps.

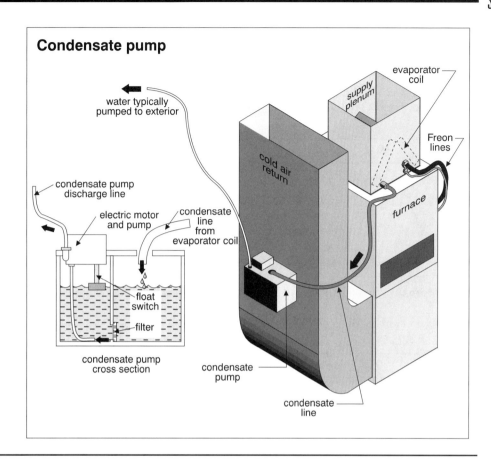

9.4.2 CONDITIONS

Common pump problems include the following:

1. Inoperative
2. Leaking
3. Poor wiring

9.4.2.1 Inoperative

Causes   Pump failure is often caused by a clogged pump or filter. Failure may also be a result of a burned out motor, seized pump or an interruption in the electrical supply.

Implication   Since the pump is located outside the evaporator coil housing and duct system, the implication of pump failure is water spilling onto the area around the pump and reservoir.

Strategy   If the system is operating, you can look for water discharging from the condensate line and listen for the pump working.

Beyond The Standards   If the system is not running, you can pour water into the pump reservoir to activate the pump. This goes beyond the Standards and many inspectors won't do this. However, if you do this test, the pump should start as the reservoir fills.

If the pump does not work, the pump may be inoperative, although you have to be careful. The electrical power to the pump may have been shut off when the air conditioner was shut off for the season. If you're looking at an air conditioning system outside of the cooling season, the condensate pump may not be powered. Be careful not to describe the pump as inoperative under these circumstances.

9.4.2.2 Leaking

The pump reservoir may be leaking. This can be caused by –

Causes
- rust
- cracking
- mechanical damage
- defective float

Implications   Water damage and a burned out pump are the implications.

Strategy   Check the pump and the area around it.

9.4.2.3 POOR WIRING

In the Electrical Module, we discuss electrical conditions that you might come across in a home. Many of these problems can be found at the condensate pump. Because of the proximity of the water and electricity, condensate pumps should be grounded. Many are not. Other common electrical conditions include loose connections, poorly supported wiring, open connection boxes and overfusing.

Causes   The cause of inappropriate wiring is usually original installation.

Implications   The implications can be fire or electrical shock.

Strategy   Check the electrical supply to the condensate pump as you check the rest of the electrical system. Ensure that the pump is grounded.

▶ 10.0 REFRIGERANT LINES

10.1 INTRODUCTION

Function And Material   The lines that carry the refrigerant between the evaporator and condenser coils and through the compressor and expansion device are typically copper.

Suction Line   The larger line typically carries a cool gas and is insulated. This is referred to as the **suction line**. It is also called the **return line**.

Liquid Line   The smaller uninsulated line typically carries a warm liquid. It is most often called the **liquid line**. Refrigerant line sets sometimes come in precharged standard lengths with fittings on either end. These sets do not have to be field cut or charged.

Allow Oil To Flow Back To Condenser   Where the evaporator coil is higher than the condensing unit, the suction line should slope down toward the condensing unit with a slope of at least one quareter inch per foot.

SECTION ONE: AIR CONDITIONING

Line Coiled Horizontally
Extra lines are usually coiled near the evaporator coil. The line should be coiled horizontally rather than vertically. The coils should allow oil to flow down through the coil and back to the condensing unit.

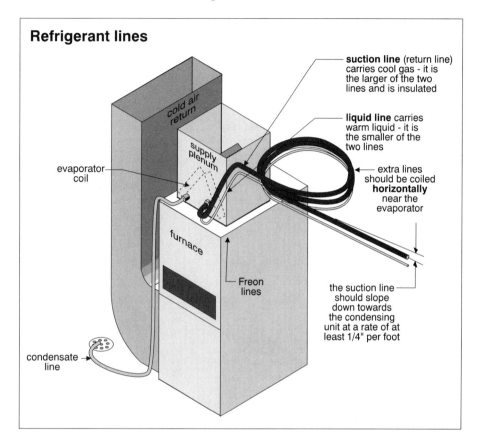

Boiling Points Of Freon 12
Pressures in the lines may range from 50 psi up to 275 psi. Since Freon is a gas at atmospheric temperature and pressure, leakage through the lines is not like a water leak. The Freon will dissipate as a gas. It may leave an oil residue. Freon 12 has a boiling point of around -20°F at atmospheric pressure. At 55 psi, its boiling point is about 30°F, and at 75 psi, the boiling point is around 40°F.

Filter/Dryer   In some installations, you will find a **filter/dryer** in the liquid line. Filter/dryers clean and dry the refrigerant. They are often added to a system where the compressor has been replaced. They help remove any contaminants. They are roughly the size and shape of a soft drink can. They may be located in the liquid line near the condenser outlet or near the expansion device.

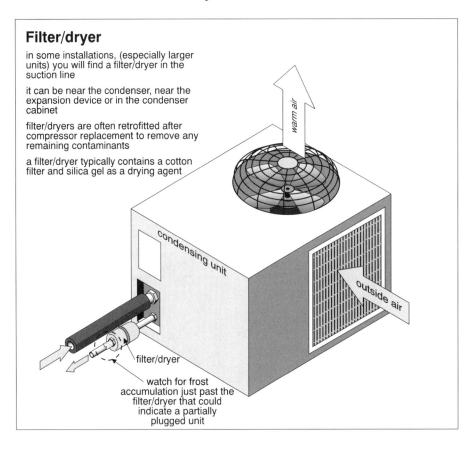

Watch For Frost   Frost accumulation just past the filter/dryer indicates a partially plugged unit and service should be recommended.

Supports and Radius Of Bends   Supports for refrigerant lines should be every 5 to 6 feet. Bends in refrigerant lines should have a minimum 12-inch radius.

Sight Glass   A **sight glass** may be installed on the liquid line, usually near the condenser. This allows the service person to check refrigerant levels. If bubbling is noted in the sight glass, this indicates possible problems and service should be recommended. The sight glass is about one inch in diameter. Many sight glasses have a colored ring. If the ring color changes, this indicates moisture in the refrigerant. This is a serious condition. About one tablespoon of moisture in the refrigerant system will destroy a compressor in a few months.

Sight glasses are more common on commercial systems than on home air conditioning systems.

SECTION ONE: AIR CONDITIONING

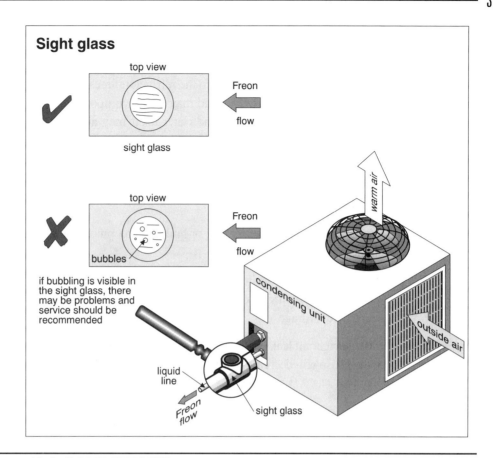

10.2 CONDITIONS

Common refrigerant line problems include the following:

1. Leaking
2. Damage
3. Missing insulation
4. Lines too warm or cold
5. Lines touching each other

10.2.1 LEAKING

A leak in a line is usually identified by the oil residue on the line. Because oil travels through the system with the refrigerant, a leak will often show up as an oil stain. Escaping refrigerant boils off and leaves no trace, other than the oil which is left behind.

Causes Refrigerant leakage is usually the result of –

- corrosion, or
- mechanical damage

SECTION ONE: AIR CONDITIONING

Corrosion And Sludge — Corrosion is often the result of contaminants in the system, copper lines touching galvanized ductwork or other dissimilar metals, or a corrosive atmosphere in the house caused by chemical storage, furniture refinishing, etc. Common refrigerants contain chlorine. If water gets into the refrigerant lines, the water and chlorine may make hydrochloric acid. The acid may attack components, and the acid will mix with the oil, creating sludge. The water itself may also create sludge if it mixes with the oil.

Mechanical Damage — Mechanical damage can occur at the outdoor unit if the unit is bumped by a lawn mower, for example. Mechanical damage may also be possible where the lines go through the exterior house wall.

Settlement — Settlement of the condenser coil or building can put considerable pressure on the copper lines.

People playing or working around the air conditioner may step on the lines, or the interior lines running through the house can be damaged by driven nails, careless handling of storage, etc.

Implications — If the refrigerant leaks out, the system performance will deteriorate, comfort levels in the house will decrease. Ultimately, the system will shut down or the compressor may fail.

Strategy — Since refrigerant line joints are usually only at the coils, concentrate on the connections to the coils at either end, where visible. If there is a thermostatic expansion valve, check its connections.

Check along the length of the line for evidence of mechanical damage, particularly in exposed areas and at the interior and exterior penetration points through the house wall. Refrigerant lines need support so that joints aren't stressed under the weight of the lines.

On attic units, pay attention to where the lines disappear into walls or ceilings.

Lines Through The Wall — Where refrigerant lines go through walls, the hole in the wall should be considerably larger than the refrigerant lines and the lines should be in conduit where they go through the wall. The ends of the conduit should be sealed with a flexible material at either end to allow movement but to prevent moisture and insect entry into the building and heat loss out of the building.

Oil Stains — Look for evidence of oil stains on the refrigerant lines. Service people use leak detection fluids, halide torches or electronic leak detectors. These are beyond our scope.

SECTION ONE: AIR CONDITIONING

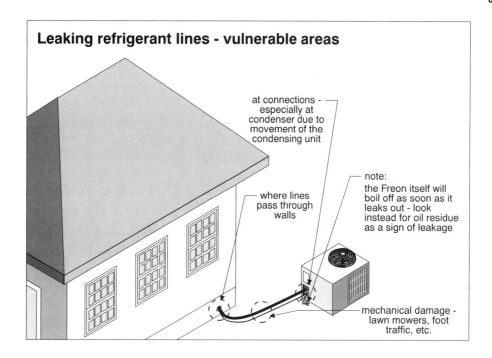

10.2.2 DAMAGE

In some cases, the lines may be crimped or damaged without leakage. Any reduction in the diameter of the lines will act as a pressure restriction, changing the temperature of the refrigerant that passes through the restricted area. This will adversely affect system performance.

Causes This is caused by impact or building settlement.

Implications Reduced performance of the system is the implication.

Strategy Follow the refrigerant lines, looking for evidence of crimping or flattening, particularly at changes in direction or where the lines go through walls or ceilings.

10.2.3 MISSING INSULATION

The large-diameter line (suction line or gas line) should be insulated along its entire length. The insulation performs two functions.

1. It prevents the suction line from sweating and dripping water inside the house.
2. The insulation also prevents the suction line attracting heat from the outdoors on its way to the condenser coil. We are trying to take heat out of the house and dump it outside. We don't want to gather outside heat and dump it into the suction line before it goes into the compressor.

SECTION ONE: AIR CONDITIONING

Causes Insulation may be missing because –

- it was never put on in the first place
- animals have chewed it
- service people or home inspectors have pulled it away to inspect or repair areas
- the insulation may have worn away

Implications System performance will be adversely affected if the outdoor insulation is missing. If the indoor insulation is missing, water damage to the home may result.

Strategy Look for the insulation to be intact along its entire length. Pay particular attention to the outdoor section of the suction line.

10.2.4 LINES TOO WARM OR TOO COLD

Cold Suction Line When the system is operating, the large insulated suction line should be cold to the touch and sweating at any point where there is no insulation.

Warm Liquid Line The smaller uninsulated liquid line should be warm to the touch after the system has been operating for 10 or 15 minutes.

Causes There are many causes for poor system performance. Some of them have been touched on earlier in this Module. You won't be to determine the causes in most cases. Further investigation should be recommended.

Frost Frost on the suction line may indicate too much refrigerant going through the expansion device, an inoperative house air fan, too much refrigerant or too low an outdoor temperature. Frost on the liquid line may mean the dryer is clogged. Frost on the expansion device may mean the device is clogged with ice.

Implications A lack of adequate cooling in the house and possible compressor damage are the implications.

Strategy Touching the lines is a part of any inspection when the air conditioner is operating. Look for the suction line to be at roughly 45°F to 55°F, and the liquid line to be at about 90°F to 110°F. Measurement with instruments is not needed.

If the suction line isn't cold, or the liquid line isn't warm, servicing should be recommended.

10.2.5 LINES TOUCHING EACH OTHER

The liquid and suction lines should not be in contact. If they are touching each other, it may be because of –

Causes
- poor installation
- building settlement
- mechanical forces

Implications Heat transfer between the lines and reduced system efficiency are the implications.

Strategy Make sure these lines don't touch each other.

SECTION ONE: AIR CONDITIONING

Air Conditioning & Heat Pumps
MODULE

QUICK QUIZ 3

☑ INSTRUCTIONS

- You should finish Study Session 3 before doing this Quiz.
- Write your answers in the spaces provided.
- Check your answers against ours at the end of the Section.
- If you have trouble with the Quiz, reread the Study Session and try the Quiz again.
- If you did well, it's time for Study Session 4.

1. The condenser coil in a split system is typically located_____

2. The evaporator coil in a split system is typically located _____

3. The condenser fan in a split system is located _____

4. What is the function of a receiver in a condenser?
   _____

5. List four condenser problems.
   _____
   _____
   _____
   _____

SECTION ONE: AIR CONDITIONING

6. Explain in two sentences the differences between a conventional air-to-air condenser and a water-cooled condenser coil.

   _____

   _____

7. List three common water-cooled coil problems.

   _____

   _____

   _____

8. List three common configurations of evaporator coils.

   _____

   _____

   _____

9. Evaporator coils should be upstream of furnace heat exchangers.
   True ☐   False ☐

10. List six common evaporator problems.

    _____

    _____

    _____

    _____

    _____

    _____

11. In one sentence describe the function of a capillary tube.

    _____

12. In one sentence describe the function of a thermostatic valve.

    _____

13. List four common expansion device problems.

    _____

    _____

    _____

    _____

SECTION ONE: AIR CONDITIONING

14. List five possible causes of leaking condensate drain pans.

15. When is an auxiliary condensate drain pan needed?

16. What are condensate drain lines usually made of?

17. Do condensate drain lines ever have a trap?

18. If so, what is its purpose?

19. List two acceptable discharge points for condensate drain lines and one generally unacceptable discharge point.

20. List three common condensate pump problems.

21. Refrigerant lines are usually made of _____

22. The larger line contains a liquid.
True ☐ False ☐

23. The smaller line is called the suction line.
True ☐ False ☐

SECTION ONE: AIR CONDITIONING

24. The larger line is called the return line.
True ☐   False ☐

25. List five common refrigerant line problems.

_____
_____
_____
_____
_____

26. How would you normally identify a leak in a refrigerant line visually?
_____

27. Which line should be cold when the system is operating?
_____

28. Which line should be warm when the system is operating?
_____

29. Which line should be insulated?
_____

If you didn't have any trouble with this Quiz, then you are ready for Study Session 4.

Key Words:
- *Condenser coil*
- *Receiver*
- *Clothes dryer vent*
- *Water cooled condenser coil*
- *Backflow preventer*
- *A-coil*
- *Expansion device*
- *Capillary tube*
- *Thermostatic expansion valve*
- *Condensate drain pan*
- *Trap*
- *Condensate pump*
- *Suction line*
- *Liquid line*
- *Freon*
- *Filter/dryer*
- *Sight glass*
- *Sludge*

SECTION ONE: AIR CONDITIONING

Air Conditioning & Heat Pumps
MODULE

STUDY SESSION 4

1. You should have finished Study Session 3 and Quick Quiz 3 before starting this Study Session.

2. This Study Session deals with the condenser fan, the evaporator fan and the duct system.

3. At the end of this Study Session, you should be able to –

    - describe the location and function of the condenser fan and the evaporator fan
    - list four common condenser fan problems
    - list seven common evaporator fan problems
    - list eight common duct problems

4. This Study Session should take you about one hour to complete.

5. Quick Quiz 4 is included at the end of this Study Session. Answers may be written in your book.

Key Words:
- *Condenser fan*
- *Evaporator fan*
- *Belt drive*
- *Direct drive*
- *Supply ducts*
- *Return ducts*
- *High returns*
- *Low returns*
- *Flex duct*
- *Obstructed registers*
- *Humidifier duct damper*
- *Duct insulation*

SECTION ONE: AIR CONDITIONING

▶ 11.0 CONDENSER FAN (OUTDOOR FAN)

11.1 INTRODUCTION

Function This fan blows outdoor air across the hot condenser coil, cooling the hot Freon gas. The gas inside the coil condenses as it cools and the air passing over the coil heats up. This is how we dump the house heat into the outdoor air.

Location And Orientation The outdoor fan is located in the condenser cabinet. The fan may rotate vertically diagonally or horizontally, depending on the manufacturer. The majority of modern units are horizontal and discharge through the top of the unit. The fan draws air in through the sides of the condenser coil and discharges the warmer air through the top of the cabinet.

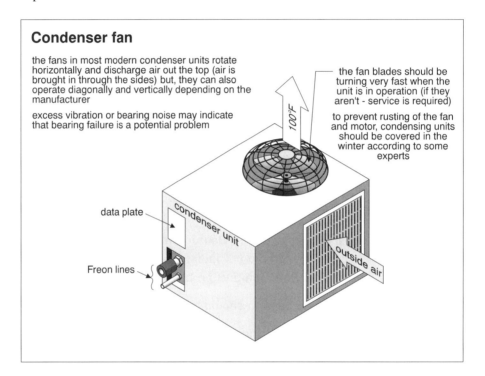

11.2 CONDITIONS

Common fan problems include the following:

1. Excess noise/vibration
2. Inoperative
3. Corrosion or mechanical damage
4. Obstructed airflow

SECTION ONE: AIR CONDITIONING

11.2.1 EXCESS NOISE/VIBRATION

Causes The two common causes of noise from condenser fans are bearing noise and fan blades that are out of balance. There may also be a foreign object interfering with the fan.

Implications Bearing noise is an indicator of a bearing about to fail, usually from a lack of lubrication.

Unbalanced fan blades will rotate but will eventually cause bearing failure. Unbalanced fan blades are usually a result of mechanical damage or dirt and will not move as much air as they should.

In any case, system failure is the ultimate result of a noisy fan.

Strategy Listen for high pitched intermittent or steady squeals from the fan area. Observe the fan blades for vibration while rotating and listen for a helicopter-type noise from unbalanced fan blades.

When the fan is not operating, look at the blades to see if they are caked with dirt or are obviously damaged or misaligned.

11.2.2 INOPERATIVE

Causes The fan may be inoperative because of an interrupted power supply. If one leg of the 240-volt power supply is disconnected, the fan will not work. (Most fans are 240-volt. Some older fans are 120-volt.)

The fan motor may fail and, in some cases, the compressor may be working but the fan may not be. The fan motor may fail as a result of a number of things, including bearing problems and obstructions.

Implications When the outdoor fan doesn't work, very little heat will be transferred to the outdoor air, which means that the air conditioning will not perform properly. Over time, compressor failure may result.

Strategy When the unit is running, ensure the fan blades are turning. Most modern fans are axial and turn at approximately 1,800 RPM. Newer quiet **low rpm fans** may turn at 875 rpm. If the fan is not moving or is moving very slowly, service is required.

11.2.3 CORROSION OR MECHANICAL DAMAGE

Causes Corrosion (rusting) of the outdoor fan is usually the result of inadequate cover during the winter months. Many fans are aluminum and do not show signs of corrosion. Covers may not be used, since no rust is visible. The motor concealed below the fan blades may rust, however.

Mechanical damage is not common, since there is grille work protecting the fan. Twigs or branches can fall into the fan area and damage a fan blade or prevent the fan from turning. Snow and ice also bend fan blades.

Implications Rusted or damaged fans may not operate. The implications are the same as for an inoperative fan.

In other cases, the fan may turn but with reduced efficiency or speed. This will move less air across the coil and reduce the heat transfer to the exterior air. This, of course, will also reduce the comfort in the house and make the whole system work harder.

Strategy Look closely at the fan and motor for evidence of rusting. A flashlight may be necessary here. Look also for evidence of mechanical damage to the motor casing and fan blades. Check for cracks in the fan blades.

The wisdom of covering condenser units during the winter months is not universally agreed on. Some say they should be covered; others say they shouldn't. People do agreed that if a cover is used, it should be breathable, so it won't trap condensation in the unit.

11.2.4 OBSTRUCTED AIRFLOW

Intake air should be unobstructed for at 1 to 3 feet adjacent to the unit (depending on the manufacturer's recommendations). The clearance on the exit side or top should be 4 to 6 feet.

The airflow through the outdoor fan may be blocked on the intake or exhaust side. A clogged, damaged or corroded condenser coil can block the air supply to the fan.

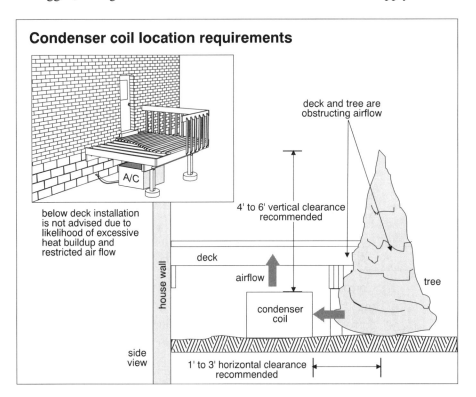

SECTION ONE: AIR CONDITIONING

Causes
- A cover left on inadvertently will obviously block both the intake and discharge.
- A tree branch on the top of the unit will obstruct the discharge.
- Condensers under decks often have obstructed flow.
- Shrubs, plants, fences and walls that are too close will obstruct the airflow.

Implications Partially or totally restricted fans will lead to non-performance and possible compressor failure.

Strategy Look at the airflow into and out of the fan.

Sunlight Or Shade Some people maintain that air conditioning condensers should be kept out of direct sunlight. Many manufacturers do not consider this a significant issue. Incidentally, manufacturers of heat pumps consider it an asset if the unit is located in the sun.

Discharge Air Is Hot When checking the operation of the outdoor fan, the air coming off the fan should be warmer than the ambient air temperature, even on a hot summer day. This is one indication that the system is operating properly.

▶ 12.0 EVAPORATOR FAN (INDOOR FAN, PLENUM FAN, BLOWER OR AIR HANDLER)

12.1 INTRODUCTION

Function The indoor fan blows house air across the evaporator coil, cooling and dehumidifing the house air.

Furnace Fan Or Fan Coil When an air conditioner is used in conjunction with a forced-air heating system, the furnace fan is the indoor fan for the evaporator as well. In independent systems, a separate fan coil unit is used, with a fan dedicated to the air conditioning system. In either case, the function is the same.

Multi-Speed Fans Where the air conditioning fan is also the furnace fan, two-speed fans are often used. The cooler, higher density air is more difficult to push through the system and, consequently, more fan capacity is required in the summer than in the heating season. Also, more airflow is needed in the summer than in the winter.

Too Much Air The size of the air handling system has to match the cooling capacity of the air conditioning equipment. If too much air is blown across a coil (the air moves by the coil too quickly), there may not be enough cooling to dehumidify the air. Noise problems will also develop and moisture will be blown off the coil, rusting ducts and damaging finishes downstream. A maximum air velocity of 500 linear feet per minute is typically recommended.

Too Little Air If the air moves too slowly, the air may be cooled dramatically, but the air may not move quickly enough through the house to achieve good cooling. Another possible impact of inadequate air flow is icing up of the evaporator coil because the coil sees a colder environment than it should.

SECTION ONE: AIR CONDITIONING

Need Twice As Much Air For Cooling

Too little air movement is a more common problem than too much air movement is. While inadequate air movement may be caused by several things, poor duct balancing, undersized ducts, dirty filters or fan problems, in this section we will concentrate on the fan performance. Roughly twice as much air movement is needed for cooling as heating with a conventional gas furnace, for example. A commonly used number is 400 to 450 CFM (cubic feet per minute) for every ton of air conditioning.

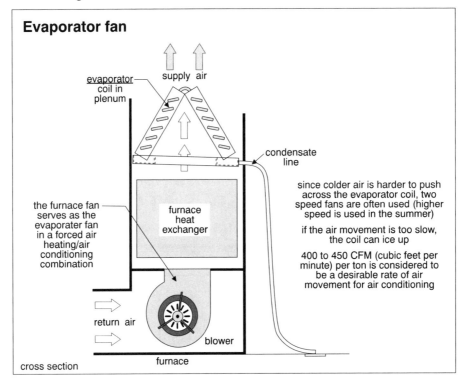

Belt Drive And Direct Drive Fan Motors

There are two common types of blowers. One uses an external motor and a **belt and pulley** arrangement to drive the fan. The other is a **direct drive** squirrel cage fan with the motor in the center of the fan. Both systems are acceptable and have their advantages and disadvantages.

Fan Coils Hung From Rafters

Fans and coils in attics are often hung from rafters rather than set on ceiling joists to minimize vibration and noise transferred to the living spaces.

Noisy air handlers or air handlers with a great deal of vibration are usually nearing the end of their life. In some cases, lubrication can restore bearing operation and eliminate noise.

SECTION ONE: AIR CONDITIONING

12.2 CONDITIONS

Common fan problems include the following:

1. Undersized blower or motor
2. Misadjustment of belt or pulley
3. Excess noise/vibration
4. Dirty fan
5. Dirty or missing filter
6. Inoperative
7. Corrosion/damage

12.2.1 UNDERSIZED BLOWER OR MOTOR

You will not be measuring the rate of airflow across the evaporator coil as part of your inspection, but you should understand that cooling air requirements are greater than most heating system air requirements, especially in the northern half of the USA and in Canada. (Note: Most modern high-efficiency furnaces require much more air movement than older furnaces.)

Causes The most common cause of an undersized fan/motor is the addition of central air conditioning to an existing forced-air heating system. Installers do not always recognize that –

- the duct system is smaller than it should be for air conditioning
- the addition of the evaporator coil into the air stream creates more resistance to airflow
- the cool dense air moved during the air conditioning season is heavier and more difficult to move than the warm air in winter
- the supply and return registers may not be ideally located
- a greater volume of air should be moved for cooling than heating

Implications Inadequate airflow means poor comfort levels in the house and may lead to icing of the evaporator coil.

Strategy You may be able to check the size of the fan motor on the furnace data plate. The data plate may specify the size of motor needed to push air at one-quarter and at half-inch static pressure (SP). The half-inch SP motor size contemplates an air conditioning coil; the one-quarter inch SP is without an air conditioning coil.

If you compare the blower motor requirements on the furnace data plate with the actual motor installed, you'll be able to determine whether the fan motor is sized appropriately, at least within the manufacturer's parameters. The motor data plate may not be accessible, particularly on direct drive blowers.

Direct Drive Blowers

On direct-drive motors, there are often four speeds at which the blower can run. The wires coming out of the motor are color-coded, and the installer can connect to various wires to generate various speeds. The black wire is always the fastest speed. If there is air conditioning, the motor should be powered through the black wire. Some units can operate on multi-speeds, so there may be a number of wires connected.

Check Airflow At Registers

When the system is operating, you will be checking the airflow at all the registers. While the use of instruments is not required, you should be able to sense reduced or no airflow at registers, which may indicate an undersized fan or a duct problem.

12.2.2 MISADJUSTMENT OF BELT OR PULLEY

On belt-driven fans, the belt may be too loose or too tight. If the pulleys don't line up properly, the fan won't work well.

Causes
- The belt often loosens as it stretches and cracks as it wears.
- The belt also may be misadjusted if the pulley sheaves (sides) are misaligned or not tightly secured to their shafts.

Implications

Misaligned or loose belts result in the blower moving less air and cause noise and vibration. Belts will fail prematurely and the stress on blower and motor bearings may shorten the life of these components.

Belts that are too tight may damage bearings on the motor and fan pulleys. The belts may also fail prematurely.

Strategy

Remove the blower door and ensure the blower cannot come on (i.e., turn off the power). Sight along the pulleys. The pulleys should be in line. Press the belt near the midpoint. With normal thumb pressure, there should be roughly one inch displacement of the belt.

Items such as loose, worn and misaligned belts and dirty fans are normal maintenance issues and are certainly no reason for people to change their mind about buying a home. However, in some cases, you will be able to attribute improper air conditioning performance to minor issues, rather than major issues such as an undersized air conditioner or an inadequate duct system. The performance of an air conditioning system can be very poor simply because of a loose fan belt, for example.

SECTION ONE: AIR CONDITIONING

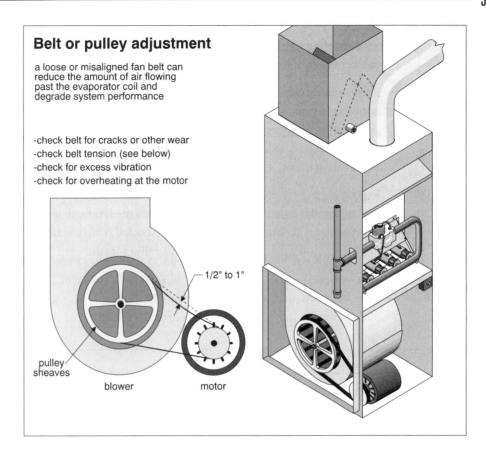

12.2.3 EXCESS NOISE/VIBRATION

Noisy air handlers generally transmit their noise through the house and annoy occupants. Noise and vibration may also indicate a failure of the fan or motor.

Causes Noisy air handlers may be the result of –

- poor adjustment or alignment
- worn bearings
- dirt on the fan blades
- a motor or fan which has come loose from its mounts,
- a foreign object caught in the air handler.

Implications Noise and vibration indicate that the fan and/or motor are not turning as freely as they should and premature failure of the air handling system should be anticipated, unless the problem is identified and corrected.

Strategy When the system is operating, listen to the fan for abnormal noises, such as squealing or grinding. These usually indicate bearing problems. Watch the blower for vibration. This may indicate that the blower is not well secured to its mounts or that there is a foreign object in the fan's path or on the blade.

Dirt On Fan Blades	Check the fan blades for dirt. Fan blades are usually slightly cupped. Dirt accumulating on the blades dramatically reduces the amount of airflow and may cause the fan to be unbalanced. This will generate noise and vibration as well.
Lubrication	Some motors require lubrication; others are permanently lubricated. While you do not want to be a service technician, sometimes noisy motors and fans simply need oil. If you see oil ports, you can mention this to your client, although professional servicing should still be recommended. Even if oil is added to stop the noise, some bearing damage will probably have taken place already.
Attic Units	Fan coils in attics can be very noisy, especially if they rest on the ceiling joists. For attic units, look for vibration isolation of the fan from the house, with hanging damper systems where the fan coil is suspended from the rafters, or rubber isolating pads below the unit if it rests on the ceiling joists.
Return Grille Location	Attic-mounted units usually have a ceiling-mounted air return grille. There should be an offset between the return grille and the fan coil. At least 5 feet of horizontal ducting helps reduce noise transmission into the house.

12.2.4 DIRTY FAN

An air handler that is caked with dust and lint will likely be unbalanced and will be subject to overheating.

Causes	The most common cause of a dirty fan is a dirty air filter.

Other causes include –

• missing air filters
• running the system with the fan compartment cover removed |
| **Implications** | The implications of a dirty air handler are an unbalanced fan (which may wear excessively and fail prematurely, possible overheating) and dirty air moving throughout the ducts and the house itself. |
| **Strategy** | When examining the air handler, look for dirt caked on the fan blades and motor. Recommend servicing, as necessary. |

12.2.5 DIRTY OR MISSING FILTERS

Whether the filters are mechanical or electronic, they need cleaning or replacement on a regular basis. Many people check and clean their filters during the winter months, but with central air conditioning, the filters must be checked year round. This is commonly forgotten.

Causes	• Dirty air filters are usually the result of lack of maintenance.
• They can also be the result of renovation activities or home hobbies, such as woodworking.

SECTION ONE: AIR CONDITIONING

Implications   The implications of dirty air filters are poor house comfort, high energy costs, premature failure of heating and air conditioning components and dirt in the ducts and throughout the house. There is very little maintenance for homeowners to do on air conditioning systems. It's amazing how often the simple act of cleaning or changing air filters is neglected.

Strategy   When testing an air conditioning system, check the air filter before running the system. If the filter is completely obstructed, poor performance may be expected. This may mask other problems. The troubleshooting of multiple and layered problems on any mechanical system is beyond our scope. Many home inspectors explain to the client the limitations of testing a system with a dirty filter. It is very difficult to know whether the poor performance is due **only** to the dirty filter.

The filters can be difficult to access on attic systems although they can often be changed from the room below by access through the ceiling return grille. They are often neglected.

12.2.6 INOPERATIVE

The fan that moves house air across the evaporator coil may not work.

Causes   The fan may not work because of –

- electrical supply problems
- a seized or obstructed fan
- a seized motor
- a broken belt or pulley
- a control problem

Implications   If we can't move air across the coil, we can't cool the house. The coil is likely to ice up, and we won't move any air through the duct system.

Strategy   This one is easy to inspect for. When the air conditioning compressor is running, the indoor fan should be running. If it is not, shut off the air conditioner and recommend servicing.

12.2.7 CORROSION/MECHANICAL DAMAGE

The evaporator fan may work, but show evidence of corrosion or mechanical damage.

Causes   Mechanical damage is usually a result of careless handling by a service person or owner. This is not common.

Corrosion may be the result of –

- flooding in the room
- chronic dampness in the area
- leakage of the condensate system for the evaporate coil
- humidifier leakage
- plumbing system leakage
- The use or storage of corrosive chemicals near by (e.g., swimming pool chemicals)

SECTION ONE: AIR CONDITIONING

Implications A damaged or corroded fan may fail prematurely. It may also not run properly, resulting in poor airflow, poor cooling, high energy costs and possible icing of the evaporator coil.

Strategy Look at the fan for evidence of damage or corrosion. Remember to shut the power off before getting in there. Remember to turn it back on when you're finished!

▶ 13.0 DUCT SYSTEM

13.1 SUPPLY AND RETURN DUCTS

Function The duct system moves cooled dry air through the house and brings warm moist air back to the evaporator coil for cooling and drying. A good duct system keeps all parts of the home at roughly the same temperature.

An air conditioning system relies as much on a good duct system as it does on good cooling equipment. Many air conditioning systems dramatically under-perform as a result of duct problems.

Vibration Collars While not required, canvas or neoprene vibration-damping collars that isolate the house ducts from the furnace or fan coil help reduce noise transmitted through the home by the duct system.

13.1.1 CONDITIONS

Common duct problems include the following:

1. Undersized or incomplete
2. Supply and return register problems
3. Dirty
4. Disconnected or leaking
5. Obstructed or collapsed
6. Poor support
7. Poor balance
8. Humidifier damper missing

13.1.1.1 Undersized or incomplete

Undersized ducts are very common, particularly where air conditioning has been added to an existing forced-air system.

Causes Undersized ducts are an installation issue.

Implications Inadequate or uneven cooling.

SECTION ONE: AIR CONDITIONING

Strategy When operating the system, check for uniformity of airflow at supply registers. Check the operation of return air registers. A piece of tissue is often needed to check that return registers are drawing properly. Check for a balance between supply and return registers. An absence of return registers or inadequately sized return registers are very common problems.

Flex Ducts Restrict Airflow Check for the extensive use of flexible ducts. Flexible ducts have far more friction loss than rigid ducts. Flex ducts can be used for branches, but should never be used for main feed ducts.

A detailed duct system analysis is beyond the scope of a home inspection, but poor system performance on an adequately sized air conditioner can often be traced to an undersized distribution system.

All Of Home Served? Be very careful to ensure that the duct system feeds all parts of the home. Additions and renovations are sometimes not air conditioned. Watch for electric heat in a room. This may mean no ductwork and, consequently, no cooling. It is very expensive to add air conditioning to spaces that don't have ducts. You don't want to have to provide a client with air conditioning in rooms because you mistakenly told them that the entire house was air conditioned.

Watch for houses with hot water heat that have had independent air conditioning systems added. These systems often don't cover the whole house.

High Level Return Grilles Retrofitted air conditioning systems have different needs than heating systems. Heating systems don't usually require a great deal of return air from the upper floors. Air conditioning systems do need good air return from upper levels.

It is common to find large temperature differences between the first and second floors on air conditioning systems that have been added to ducts designed for heating only.

Rough Guidelines If you are suspicious about the size of the duct system, there are some rough approximations you can do to get a sense of the adequacy. First, measure the cross-sectional area of the supply and return plenums at the furnace or fan coil. They should be the same size.

SECTION ONE: AIR CONDITIONING

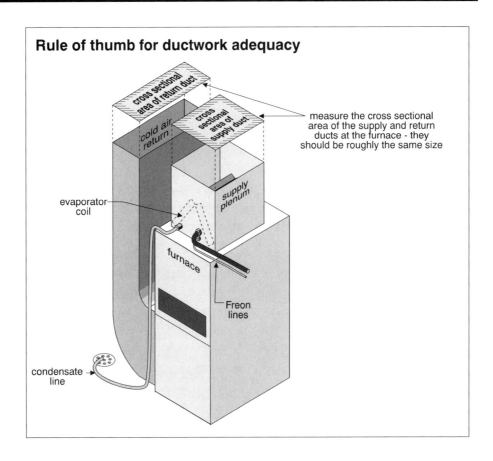

High And Low Returns

Count the individual return grilles on the different floors (count double-width grilles as two grills). In a two story home, the second floor should have at least as many returns at the first floor. Look for the second-floor returns to be at high level, or have both high and low level returns, with low level returns that can be shut off during the cooling season.

Large or medium-sized rooms with more than one exterior wall should have at least two supply registers.

Longer duct runs should be larger diameter than shorter runs.

Attic And Crawlspace Ducts Are Inefficient

It is more efficient to run ducts through conditioned spaces than attics or crawlspaces, but this is often not practical. Even with insulation and vapor barriers, some unwanted heat gain to the conditioned air will take place as the air passes through unconditioned spaces. Expect system performance to be lower if most of the ducts run through attics or crawlspaces.

Most Systems Are Compromises

None of this is conclusive, and most systems will have a design that is imperfect in some way. If you perform these rough checks in several houses, you will come to know what is typical for your area and be able to highlight unusually bad situations.

SECTION ONE: AIR CONDITIONING

13.1.1.2 Supply and return register problems

Registers are associated with duct problems, but can be looked at slightly differently. A good-sized air conditioning system and reasonable-sized ducts may still result in poor performance if registers are not of the appropriate size, number and location. Conditioned air is heavier than the room air and may be expected to fall. The warmest air is at the top of the room.

An Ideal Arrangement

Introducing conditioned air near the ceiling on an outside wall allows it to fall through the warm air, creating a relatively even temperature throughout the room. The return registers are ideally located on the opposite (interior) side of the room to enhance circulation through the entire room. The return registers should also be high to carry the warmest air out of the room back to the air conditioner to be cooled.

Many systems are not set up this way and, when the same ducts are used for heating and for cooling, compromises are usually made. Heating systems are best set up with the supply and return registers near floor level, just the opposite of what we talked about for air conditioning registers.

Supply On Outside Walls

During the summer, the warmest part of the room is likely to be the outside wall, especially if there are windows in the wall. It makes sense to introduce the coolest air close to the outside wall. Therefore, the supply register should be located near the outside wall.

Returns On Interior Walls

The return registers should be on an opposite and interior wall. In this regard, heating and air conditioning systems have the same register location goals. For the heating system as well, the supply register should be located near an outside wall (usually below windows) and the return register should be located on an opposite and interior wall.

Heating Is The Priority

Real-life compromises often mean that registers are set up with heating as the priority over air conditioning. For example, in a two-story house with the ducts designed primarily for heating needs, the return air grilles typically collect cold air that falls down stairwells to the lower parts of the home. Consequently, the return ducts and grilles are usually located centrally, near the bottom of stairwells, and there is often more air return on the lower level than the upper level.

SECTION ONE: AIR CONDITIONING

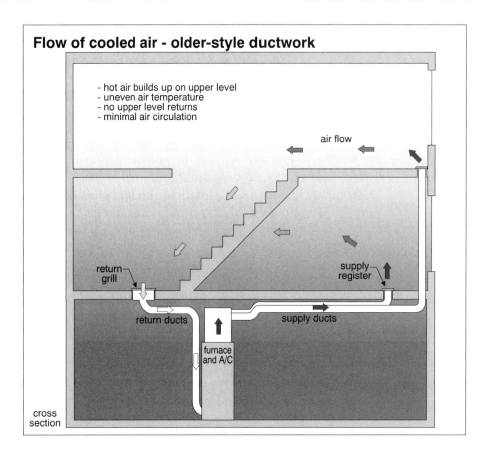

An ideal return air system for air conditioning would have more return air ducts and grilles on the upper level, and they would be located near the top of the second floor. These grilles would be centrally located and above stairwells.

SECTION ONE: AIR CONDITIONING

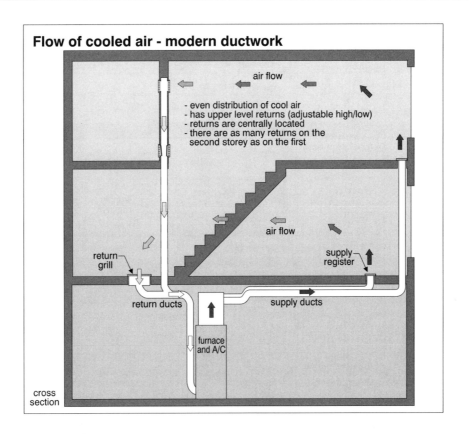

Flow of cooled air - modern ductwork
- even distribution of cool air
- has upper level returns (adjustable high/low)
- returns are centrally located
- there are as many returns on the second storey as on the first

Cooling Better On First Floor Than Second — Since most systems are set up with heating as a priority, it is very common to find houses where the air conditioning system works reasonably well on the first floor, but not nearly as well on the second floor.

Adding High Level Return Inexpensively — One of the enhancements that can be made to improve the air conditioning performance is the addition of a very inexpensive high-level air return register immediately above a low-level return register, and replacement of the low return register with one that includes a damper.

Return air registers in wood frame walls often use the space between two studs as the return duct. Adding a register near the ceiling level, immediately above the cold air return register, allows warm air to be taken from high in the room.

Damper Closed In Summer — The lower cold air return register is typically replaced with a register that has a damper that can be opened or closed. In the summer, the damper is closed so that cool air will not be drawn in through the low register and the high level register will draw warm air from the room back to the air conditioning coil.

Damper Open In Winter — In the winter, the damper on the low level register is swung open, which allows cool air to be drawn off at floor level. As the damper is opened, it blocks off the air path from the high level return register. This damper acts as a diverter; so we won't pull warm air from high in the room.

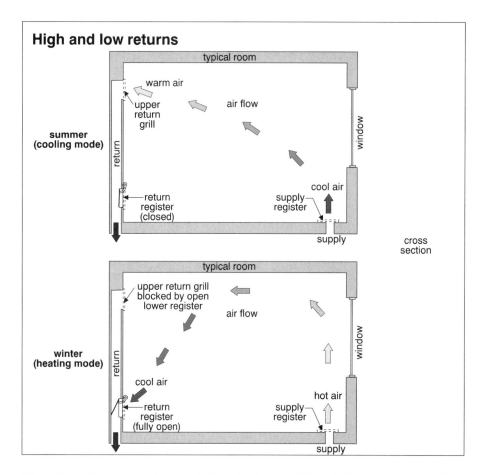

Don't Design — Many home inspectors stop short of designing specific solutions to house problems and that is wise. If you are going to suggest solutions such as the one we've just talked about, you should be very careful. You don't want to over-promise, or suggest that this relatively minor and inexpensive improvement will completely change the performance of an inadequate air conditioning system.

Obstructed Registers — One of the most common register problems is that either supply or return registers are obstructed by carpets or furniture. Performance improvement can often be achieved simply by uncovering obstructed supply and return registers.

Never Enough Registers — In the perfect home, every room would have independent supply and return registers. In the real world, this doesn't happen. As we have discussed, undercutting doors to allow air circulation is the common solution. You should, however, understand that it is a compromise. Many of your clients will have trouble understanding why, if there is an active supply register, can it be important that there be return air?

SECTION ONE: AIR CONDITIONING

Rooms With No Returns Are Dead Ends

The analogy you might give them is to think of the closed room as a balloon. You can blow air into the room until it is full of air or pressurized. If the air can't get out of this room, the fan will push the air through the duct system in other places where the air can move in a loop. You have effectively created a large dead end or balloon in this room. This air will heat up during the summer or cool during the winter, and the room will be uncomfortable.

To summarize, the conditions you're likely to find with supply and return registers are –

1. inadequate number
2. poor location
3. obstructed registers

Causes
- Inadequate number and location of registers may be an original design issue.
- Where air conditioning has been added to a heating system, it is the limitation of the existing ducts that causes the condition.
- Obstructed registers are homeowner lifestyle issues.

Implications
A well-sized and properly operating air conditioning system may still produce an uncomfortable house, if registers are not adequate in number, proper in location and free to flow.

Strategy
As you go through the house, check not only the airflow through the registers, but ensure there are registers in all rooms. Pay attention to the location of the registers. Are they located ideally or are they a compromise?

Don't forget to check the return air register flow. In many cases, a piece of tissue is needed to check that air is being drawn in through the return grilles.

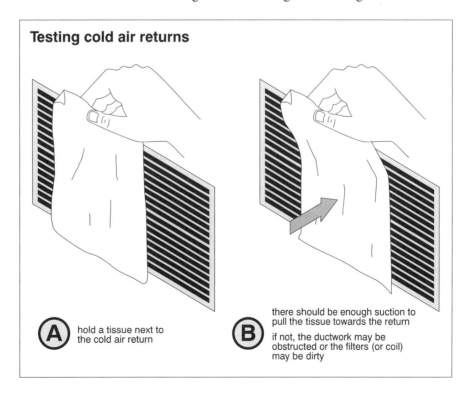

Testing cold air returns

(A) hold a tissue next to the cold air return

(B) there should be enough suction to pull the tissue towards the return

if not, the ductwork may be obstructed or the filters (or coil) may be dirty

Where there are high and low return registers with a damper on the low return, explain the operation of these to your client.

13.1.1.3 Dirty Ducts

While you can't get a look at the inside of all the ducts, you should be aware of this problem.

Causes Dirty ducts are usually the result of poor filter maintenance. Filters may be missing or they may be poorly fit, allowing air to go around the filters.

Electronic air cleaners may or may not be an advantage. Poorly maintained electronic air filters are every bit as ineffective as poorly maintained conventional filters.

Implications Reduced airflow lowers comfort levels and increases the amount of dirt in the air, with possible adverse health affects.

Strategy You can't look at the inside of the ducts in very many areas, so most of your clues will be indirect. You should be able to check the ducts in the area of the filter. The condition of the filter will be a clue as to the condition of the ducts.

If you check the evaporator coil and find it clogged, you can assume there is a fair bit of dirt in the ducts as well.

You can also check the ducts immediately adjacent to the registers for evidence of dirt. Duct cleaning services are readily available in most areas. A well-maintained system can go many, many years without having to clean the ducts. A poorly maintained system may require cleaning. The bigger issue is the cleaning of the evaporator coil, in most cases.

Evidence Of Cleaning You may be able to determine whether the duct system has been cleaned. Round plugs, roughly 1½ inch in diameter, along the ductwork and 6-inch square metal patches near the furnace or air handler indicate professional duct cleaning has been done.

13.1.1.4 Disconnected or leaking ducts

Ducts that are not continuous allow conditioned air to leak out of the system and unconditioned air to leak in. Disconnected or leaking ducts are far more important in unconditioned spaces, such as attics and crawlspaces.

Causes Disconnected or leaking ducts may be the result of improper installation. In some cases, the ducts were never connected. In other cases, the connections were not made tightly. Original connections may work loose with vibration. House renovations and improvements may disturb ducts, causing leaks.

Implications The implications are reduced house comfort, longer on-cycles and a system that has to work harder.

SECTION ONE: AIR CONDITIONING

Strategy Follow as much of the duct system as you can, preferably with the system operating. Pay particular attention to joints for air leaking out on supply ductwork and air leaking in on return ducts. Watch for mildew on return ductwork where warm, moist attic air may be drawn into the ducts and cooled to its dew point. The moist environment created here may support mold, mildew and bacteria growth.

13.1.1.5 Obstructed or collapsed ducts

Causes Ducts may be obstructed by foreign objects dropped into registers. Collapsed ducts may be the result of animals or people crawling through attics or crawlspaces.

Ducts In Floor Slabs Ducts buried in concrete floor slabs may be partially collapsed during original construction, when the concrete is poured. These ducts may also be obstructed by debris or water that accumulates in the ducts. Water accumulating in the ducts may rust metal ducts causing them to collapse, creating a substantial obstruction.

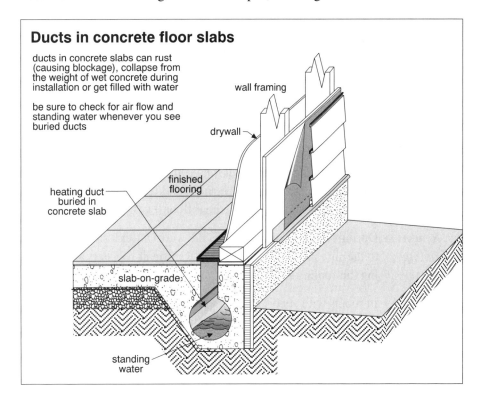

Health Issues Beyond the scope of a home inspection are the health issues associated with water in ducts in slabs. Also beyond our scope is the issue of asbestos cement ducts in floor slabs and health issues related to asbestos.

Implications Obstructed or collapsed ducts will reduce the airflow, lowering comfort levels and making the system work harder.

Strategy Trace the ducts and look for evidence of irregularities. In some cases, it is possible to remove registers and put a light and a mirror down to look at the condition of the duct supplying that register. This is particularly helpful for ducts buried in slabs.

SECTION ONE: AIR CONDITIONING

13.1.1.6 Poor support for ducts

Inadequately supported ducts may become disconnected. Flexible ducts that are not well supported will have more friction loss than they should, resulting in poor airflow.

Causes Poor support may be an original installation problem or support may have been compromised by people working in and around the system.

Implications The implications are leaks in the ducts or reduced airflow caused by sagging ducts.

Strategy As you trace the ducts, look for evidence of sags or broken or missing supports.

13.1.1.7 Poor balancing

Needs Seasonal Balancing The balance of an air conditioning system is usually different from the balance of a heating system. Modifications should be made during seasonal changes. Many homeowners do not understand or make these adjustments.

Balancing Dampers Poor balance may be, in part, an original design issue, or it may be an adjustment problem. Most branch ducts have adjustable internal dampers. Supply registers can typically be opened and closed, and many systems have a second internal damper just upstream of the register that can be adjusted by hand after removing the register.

Low Returns For Heating As we discussed earlier, some systems employ both high and low return air registers to deal with the different seasons. During the heating season, it is the cool air that should be collected from the room and returned to the furnace to be heated. The return air registers should be near floor level.

High Returns For Cooling During the cooling season, it is the warm air that should be collected and returned to the air conditioner for cooling. Return air registers are ideally at high level during the cooling season.

Undercut Doors To Allow Circulation In rooms with supply registers only, and doors that can be closed, the door should be undercut at least ¾ inch to allow air to reach return registers. The air has to circulate through the house for heating or air conditioning systems to work properly. Restriction to the circulation at any point impacts the whole system.

SECTION ONE: AIR CONDITIONING

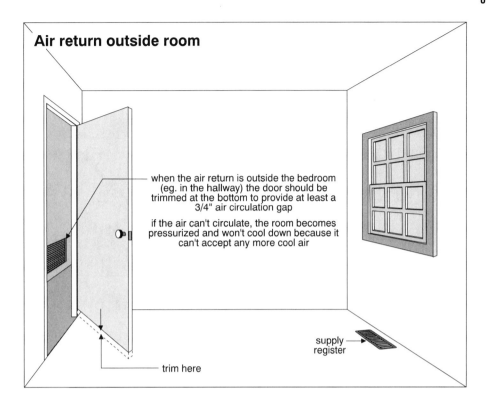

Causes	Original installation or poor set up or maintenance contribute to poor balancing.
Implications	Ineffective cooling, excessive costs and shortened life expectancy for the cooling equipment are the implications.
Strategy	As you check the airflow through the registers, look for uniformity of air movement. Check the return registers, as discussed earlier.

If no air is felt at a supply register, remove the register and make sure the balancing damper behind is not closed.

13.1.1.8 Humidifier damper missing

As discussed in **The Home Reference Book**, bypass humidifiers are common on furnaces. These humidifiers typically use 6-inch ducts that bridge the supply air plenum downstream of the furnace and the return air plenum close by. This creates some minor short-circuiting of airflow, which is not a big issue during the heating season.

Icing Evaporator Coil

This short-cycling is a problem during the cooling season, since it delivers cooler air than was intended to the evaporator coil. This can lead to icing up of the evaporator coil.

Humidifier bypass ducts should have a damper that can be closed during the cooling season to prevent this short-cycling problem. The presence of the damper should be verified. It should be appropriately marked, and your client should understand that it should be closed for the summer and opened for the winter.

SECTION ONE: AIR CONDITIONING

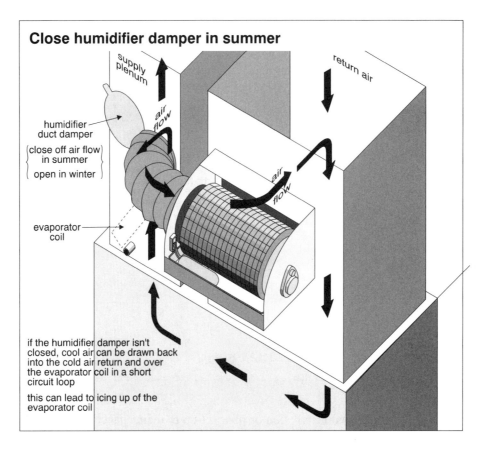

Cause	A missing humidifier duct damper is an installation mistake.
Implications	Ineffective air conditioning, iced-up evaporator coil, reduced comfort levels and damage to the compressor are all possible results.
Strategy	Check for the presence of the duct damper whenever there is a bypass humidifier and central air conditioning.

13.2 DUCT INSULATION

Ducts passing through unconditioned spaces, such as attics or crawlspaces, should be insulated and air-tight. We don't want to collect heat from these areas, or distribute cool air into these areas. As a general rule, the more ducts that pass through unconditioned spaces, the greater is the challenge for the air conditioning system.

SECTION ONE: AIR CONDITIONING

Insulation And Vapor Barriers

The insulation should have a vapor barrier and, like all vapor barriers, it should be on the warm side of the insulation. In an air conditioning environment, this means the vapor barrier should be on the outside of the insulation. The warm, moist air in the attic or crawlspace may condense if it is cooled. We want to keep the attic and crawlspace air warm, but if we inadvertently cool this air, we don't want condensation to form on the outside of the ducts. The vapor barrier is designed to prevent the warm, moist air from getting close to the ducts where it would inevitably form condensation. It also prevents moisture moving through the insulation by vapor diffusion.

The downside to all of this is that putting the vapor barrier on the outside of ductwork that runs through unfinished spaces creates a vulnerable situation. Vapor barriers are often torn.

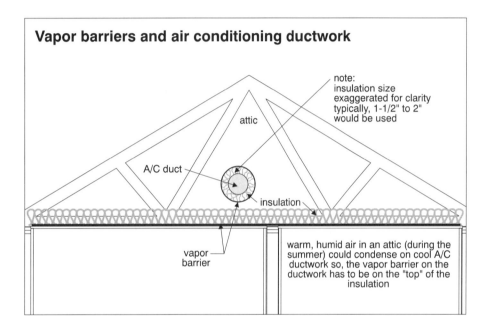

13.2.1 CONDITIONS

There is one common duct insulation problem:

1. Insulation missing or incomplete

13.2.1.1 Insulation missing or incomplete

Ducts for air conditioning runs in unconditioned spaces may be insulated when it is manufactured, or insulation can be added. The majority are pre-insulated. The insulation typically has an R value of roughly 7, but can be up to R-12.

SECTION ONE: AIR CONDITIONING

Causes
- Insulation may be missing because of an original design failure.
- The insulation may be falling off because of poor support or mechanical damage.
- In some cases, the insulation has been removed intentionally to facilitate installation in tight spaces.
- Insulation may also be incomplete because of deterioration of the insulation material over time.

Implications A lack of insulation will allow heat to be picked up by the ducts, adversely affecting the system operation. Further, the warm, moist air on the outside of the ducts may cool to a point where condensation forms. This condensation may create water damage to living spaces below and may rust metal ducts.

Condensation The implication of a missing or damaged vapor barrier is condensation forming on the ducts. Condensation forming on or in the insulation reduces the effectiveness of the insulation, and the whole system will perform poorly as a result.

Strategy Look at as much of the ductwork as you can. Look closely in tight spots to ensure that insulation is present and intact. Look for tears in the vapor barrier. Look for joints in the insulation or vapor barriers and check these closely for evidence of condensation if the system is operating.

On supply ducts, feel for cool air that may be escaping. If the duct is perforated, expensive replacement may be necessary.

Independent Retrofit Systems There are some independent air conditioning systems specifically designed to be retrofit into existing houses. These use very small diameter ducts (approximately 2 inches) wrapped with insulation and a vapor barrier.

Small Diameter Insulated Ductwork The total diameter of the duct and insulation is such that it can be fit through conventional wall stud spaces, $3^1/_2$ inches wide. These systems typically employ a large number of these small ducts and there may be several outlets in every room. The ducts and insulation are all subject to mechanical damage.

The possibility of mechanical damage to the insulation or ducts is heightened with these systems, since they are typically added to existing houses, with efforts made to minimize the disruption to finish surfaces.

These systems typically deliver high-velocity air, since the duct area is small. Like other systems, this may be noisy or drafty.

SECTION ONE: AIR CONDITIONING

Air Conditioning & Heat Pumps
MODULE

QUICK QUIZ 4

☑ INSTRUCTIONS

- You should finish Study Session 4 before doing this Quiz.
- Write your answers in the spaces provided.
- Check your answers against ours at the end of this Section.
- If you have trouble with the Quiz, reread the Study Session and try the Quiz again.
- If you did well, it's time for Study Session 5.

1. Where is the condenser fan in a split system located?

2. What is the typical inlet and outlet temperature of air across the condenser fan?

3. List four common condenser fan problems?

SECTION ONE: AIR CONDITIONING

4. Where is the evaporator fan typically located in a split system?

_____

5. What other function might the evaporator fan perform?

_____

6. List two types of drives for evaporator fans.

_____

_____

7. The amount of air that must cross an evaporator coil in a heating climate is less than the amount of air that must cross a conventional furnace heat exchanger coil.
True ☐   False ☐

8. List seven common evaporator fan problems.

_____

_____

_____

_____

_____

_____

_____

9. What are the implications of a dirty or missing air filter with respect to the evaporator fan?

_____

_____

10. Oversized ducts are a common problem.
True ☐   False ☐

SECTION ONE: AIR CONDITIONING

11. List eight common duct problems on cooling systems.

_____

_____

_____

_____

_____

_____

_____

_____

12. Flex ducts have less friction loss than rigid ductwork.
True ☐   False ☐

13. High level returns are more appropriate for heating than cooling.
True ☐   False ☐

14. Ducts passing through unconditioned spaces such as attics or crawlspaces should be

_____

15. Why should bedroom doors be undercut if there is no return air grille in the bedroom?

_____

_____

16. What clues might you find to indicate that the ductwork has been professionally cleaned?

_____

_____

_____

17. What is the purpose of the humidifier duct damper?

_____

SECTION ONE: AIR CONDITIONING

18. What is the implication of finding it in the wrong position?

_____

19. What is the wrong position in the summer?

_____

20. What is a typical level of insulation on ductwork?

_____

If you didn't have any trouble with this Quiz, then you are ready for Study Session 5.

Key Words:
- **Condenser fan**
- **Evaporator fan**
- **Belt drive**
- **Direct drive**
- **Supply ducts**
- **Return ducts**
- **High returns**
- **Low returns**
- **Flex duct**
- **Obstructed registers**
- **Humidifier duct damper**
- **Duct insulation**

SECTION ONE: AIR CONDITIONING

Air Conditioning & Heat Pumps
MODULE

STUDY SESSION 5

1. You should have finished Study Session 4 and Quick Quiz 4 before starting this Study Session.

2. This Study Session deals with thermostats, life expectancy, evaporative coolers and whole-house fans.

3. At the end of this Study Session, you should be able to –

    - describe the function and appropriate locations for thermostats
    - list seven thermostat problems
    - give the normal life expectancy for conventional air conditioner compressors
    - describe the function of evaporative coolers
    - indicate the type of climates where evaporative coolers are used
    - name three types of evaporative coolers
    - list 13 common cooler problems
    - describe the function and location of a whole-house fan
    - list two common whole-house fan problems

4. This Study Session should take you roughly one hour to complete.

5. Quick Quiz 5 is included at the end of this Study Session. Answers may be written in your book.

SECTION ONE: AIR CONDITIONING

Key Words:
- *Central thermostat*
- *Compressor life*
- *Compressor warranty*
- *Evaporative coolers*
- *Swamp coolers*
- *Single pass*
- *Rotary, spray and drip*
- *Rust*
- *Air gap*
- *Whole-house fan*
- *Louvers*
- *Backdrafting*
- *Combustion appliances*
- *Attic venting*

SECTION ONE: AIR CONDITIONING

▶ 14.0 THERMOSTATS

14.1 INTRODUCTION

Function And Location — Thermostats are designed to turn the cooling system on and off at the right times to keep us comfortable. Thermostats aren't mounted on the air conditioner. Thermostats are put in the rooms where we live and should be in a location that's representative of the average temperature.

Types — Thermostat are switches that tell the air conditioner to come on or off to maintain the house temperature at the level set by the occupant. Thermostats can be line voltage (120 volts AC or millivolt system. The low voltage thermostat is the most common and is arguably the best.

Bimetallic Strips — Traditional thermostats operate on a **bimetallic strip**. The strip made out of two different metals will bend as the temperature around it changes. In conventional thermostats, this bimetallic strip is wound into a loose coil. As the strip expands and contracts, it closes and opens a switch. These switches are typically either a mercury bulb or a snap-action switch.

Mercury Bulb Switch — In a **mercury bulb switch**, liquid mercury (which is a good electrical conductor) is at one end of a glass bulb. The glass bulb rocks back and forth like a teeter totter as the temperature rises and falls. Two wires are stuck into one end of the glass bulb, but do not touch each other.

Turning The Air Conditioner On — When the bulb tips toward the wires, the liquid mercury runs to that end of the glass bulb and submerges the ends of the wires. The electricity can flow from one wire, through the mercury, into the other wire. This completes a circuit and allows the air conditioner to come on.

Setting The Temperature — The temperature is set or adjusted by moving a needle that changes the angle of the bulb. The needle is calibrated to a gauge or dial that allows the user to select a desired temperature. This is why thermostats have to be installed level.

Snap-Action Switch — A **snap-action switch** works on a similar principle except that a magnet on the end of the bimetallic strip pulls one contact against another to close the circuit. As the metal cools and shrinks, the magnet retreats from the contacts and the switch falls open. These contacts are also often in a glass bulb, but there is no mercury in this case.

Thermostats are made of a mounting plate that contains the wiring, a bimetallic strip and setting needle. There is also a cover designed to protect the components from dust, dirt and mechanical damage. A gauge allows you to set the thermostat at a desired temperature, and a thermometer shows the present temperature. Many thermostats are designed to control both the heating and the air conditioning systems. This is a good arrangement because most of the thermostats won't allow the heating and cooling systems to be on at the same time. This would be an inefficient situation and the air conditioning system would be damaged.

SECTION ONE: AIR CONDITIONING

Sophisticated Thermostats

Sophisticated thermostats have special features:

- Set-backs allow you to have different temperatures at different times of the day and night.

- Heating and cooling functions may be performed by the same thermostat. The thermostat can control both the heating and cooling systems.

- The thermostat controls the house air fan. We can have the fan come on only with the furnace, or have the fan on all the time.

- Electronic thermostats, which have digital read-outs and lots of function, can be very flexible. An unfortunate result of this is that they can also be very complicated.

Good home inspectors are either familiar with the common programmable types of thermostats and know how to adjust them temporarily without disturbing programmed settings, or carry instructions with them for the common types. Many homeowners are intimidated by these and can't reset them if you mess up the settings. This is a good way to upset a homeowner.

Let's look at some of the common problems we'll find with thermostats, keeping in mind that we don't want to be service technicians.

14.2 CONDITIONS

Common thermostat problems include the following:

1. Inoperative
2. Poor location
3. Not level
4. Loose
5. Dirty
6. Damaged
7. Poor adjustment/calibration

14.2.1 INOPERATIVE

To activate an air conditioning system, the thermostat is turned to a lower temperature than the room temperature. If the air conditioner does not respond after about seven minutes, the system should be described as inoperative.

The thermostat is one component that may cause the system not to operate. There are many other problems that may result in the failure to respond to the thermostat. It's beyond the scope of a home inspection to troubleshoot air conditioning systems, although a bad thermostat can be included as one of the possibilities.

SECTION ONE: AIR CONDITIONING

Causes The thermostat may be malfunctioning because:

- The electrical supply is interupted.
- There is a loose connection.
- The thermostat is out of level.
- The contacts may be compromised.
- There may be a fault in the electronic circuitry.

Implications The air conditioning system won't respond if the thermostat isn't working. It may not start or it may not shut off.

Strategy Turn the thermostat down to see if the system operates. If it does not, you may remove the cover (beyond the Standards) and check that the mercury bulb switch moves properly and that the contacts are working. You may also verify that the low voltage wiring to the thermostat is properly connected. Beyond this, there isn't much you can do.

Calibration It's possible that the thermostat works, but the calibration is wrong. You may not know this during the inspection, and an analysis of the thermostat calibration is beyond the scope of your inspection.

Indoor Fan Starts Immediately The air conditioner controls are a little different from the furnace control. When the furnace is activated, it is typical for the burner to come on immediately, but the fan will not come on until the furnace is up to temperature. Air conditioning systems don't work this way. The indoor fan comes on as soon as the compressor and outdoor fan starts.

14.2.2 POOR LOCATION

Central Thermostats should be located near the center of the home. They should not be located where they will see direct sunlight or direct radiant heat from fireplaces, light bulbs, etc.

Not In Kitchens Thermostats should not be located on exterior walls. They should not be located in kitchens (since kitchens tend to be warmer than the rest of the house when people are cooking or washing dishes).

Not Against Ducts, Pipes Or Wall Ovens Thermostats should not be located on walls containing ducts or piping that may be considerably warmer or cooler than room temperature. Watch for thermostats in dining rooms, living rooms or halls that back onto wall ovens in the kitchen.

Chimneys Similarly, thermostats should not be located on walls that contain chimneys.

Away From Doors And Windows Thermostats should not be located adjacent to doorways or windows that will be opened frequently during the heating season. These drafts can cause heating systems to come on frequently for short periods.

SECTION ONE: AIR CONDITIONING

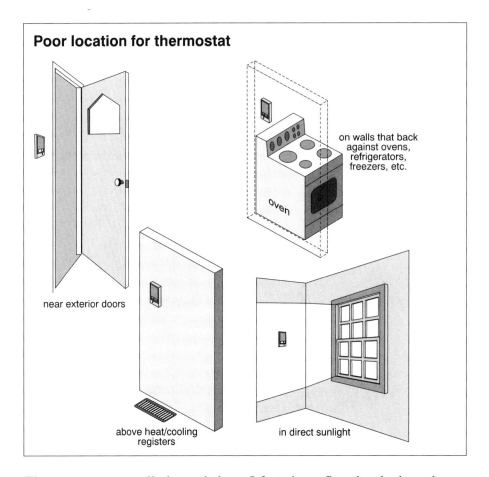

Poor location for thermostat
- near exterior doors
- on walls that back against ovens, refrigerators, freezers, etc.
- above heat/cooling registers
- in direct sunlight

Five Feet Above Floor — Thermostats are usually located about 5 feet above floor level where they are easy to read. They should not be above televisions, lamps or other heat-generating appliances.

Causes — Poor location of a thermostat is usually an installation issue, although it can be the result of house re-arrangement.

Implications — The house is likely to be less comfortable than it should, since the thermostat will think the room is either warmer or colder than it actually is. Energy bills may be higher than they should be. The system may also short cycle.

Strategy — If the thermostat is in a poor location, advise that there may be a comfort issue, and that relocating the thermostat is not a big expense.

14.2.3 NOT LEVEL

Mercury bulb thermostats need to be level to work properly.

Causes — Out-of-level thermostats that are are usually an installation error. They are often reinstalled carelessly by wallpaper hangers, amateur or professional.

Implications — The set temperature of the thermostat will not be an accurate reflection of the room temperature.

SECTION ONE: AIR CONDITIONING

Strategy With square or rectangular thermostats, it's easy to check whether the thermostat is level, within a few degrees. With a circular thermostat, it's a little harder to tell whether the thermostat is level. If you pull the cover off (which goes beyond the Standards), most thermostats have either leveling lines or leveling posts that can be checked. Some people use a small spirit level to check this, although this goes beyond the Standards, as well.

During the heating season, you'll be able to tell whether the room is roughly comfortable or not. It the room has to be set at 90°F or at 60°F to keep it at roughly 75°F, the thermostat may not be level. There could also be a calibration problem.

Remember, we're not troubleshooting these problems and don't have to be technicians.

14.2.4 LOOSE

Thermostats must be well secured to the wall in order to work properly.

Causes Thermostats may be loose because of poor installation, redecorating (including wallpaper activity that has removed the thermostat temporarily and not re-secured it well) or mechanical impact.

Implications This may result in a faulty reading on the thermostat, resulting in large temperature swings or excessive on/off cycles for the furnace.

Strategy You're going to be turning the thermostat down anyway to test the cooling system. Gently grab the thermostat base (not just the cover) and make sure it's well secured.

14.2.5 DIRTY

Causes Thermostats are often dirty as a result of a renovation project. It could be a poor maintenance condition, but this would be unusual.

Implications The thermostat may not respond accurately and eventually may not respond at all.

Strategy You can get a sense of the cleanliness of the thermostat by looking at the cover. If you remove the cover (beyond the Standards) you can also get a look at the cleanliness of the unit itself. Sometimes the cover can be wiped clean but the thermostat itself is filthy. Dirty contacts may not complete the circuit, resulting in a system that won't operate. This is a common post-renovation problem.

14.2.6 DAMAGED

Causes Thermostats may be damaged by impact or water running down the wall.

Implications Implications range from no cooling to excessive cooling with extreme temperature fluctuations and frequent on/off cycles. Poor comfort and high energy costs are common.

Strategy Look for evidence of mechanical damage. Again, if you take the cover off the thermostat (beyond the Standards) you may be able to identify concealed damage.

SECTION ONE: AIR CONDITIONING

14.2.7 POOR ADJUSTMENT/CALIBRATION

In most cases, you're not going to worry too much about the thermostat being out of calibration by one or two degrees. People will keep the house at whatever is comfortable for them. It doesn't matter a great deal if people think 74°F is comfortable based on a thermostat setting, or whether they get the same temperature and are comfortable by setting a thermostat to 71°F. However, if the room feels comfortable and the thermostat is set at 85°F, it may be so far out of calibration that the air conditioner operation is not predictable.

Implications Again, excessive temperature fluctuations may result.

Strategy Check that the temperature setting of the thermostat is close to the temperature in the house.

► 15.0 LIFE EXPECTANCY

The Standards require that you identify whether the equipment is near the end of its normal life expectancy. Although air conditioners have many components, the compressor is the heart of the unit. It is, by a wide margin, the most expensive component to replace on an air conditioner or heat pump.

Checking The Age One can determine the age of the unit from the condenser unit data plate. The **Carrier Blue Book** is helpful in this regard. Since compressors are often replaced, the home inspector who goes above and beyond will check on a unit that is more than 5 years old to determine whether the compressor has been replaced. This involves removing the cover from the outdoor condenser unit and checking the compressor itself. There are only a few manufacturers of compressors and the date coding systems can be determined by contacting the manufacturers. It's common to find older systems with relatively young compressors. If you aren't going to go this far, when you see old air conditioners, you should allow for the possibility that the compressor has been replaced.

Life Expectancy Life expectancies vary. Many people in the southern United States consider 8 to 10 years a typical compressor life expectancy. In more moderate climates, a 10-to 15-year life expectancy is typical, and in northern climates, life expectancies of 15 to 20 years may be appropriate.

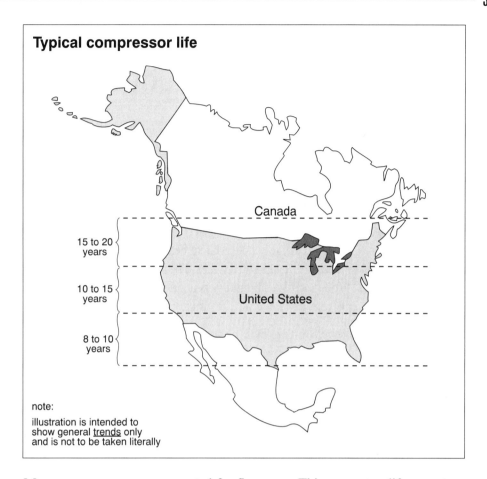

Warranty	Many compressors are warranted for five years. This suggests a life expectancy range.
Cause	The cause of old equipment is simply time.
Implication	There may be no implication with respect to existing performance. We may simply be anticipating breakdown. However, older compressors may suffer compression ratio decreases as valves wear. They may also be more expensive to operate because they draw more electricity. Some people recommend replacement of old compressors even if they are still operable, as a way to reduce operating costs through enhanced efficiency.
Strategy	The age can be determined from the data plate on the condenser or, as discussed, from the compressor itself. Interpretation of serial numbers is sometimes required to verify age. The compressors typically have a date tag riveted or glued onto the shell of the unit.
	Speak with manufacturers and installers in your area to get a sense of the common life expectancies for air conditioning compressors in your climate. If you are going to include life expectancy comments in your reports, make it clear to the client you are dealing with a probability and not a certainty. Many compressors will fail before their normal life is up and others will go well beyond what is expected.

SECTION ONE: AIR CONDITIONING

You may also ask at what age it makes more sense to replace the whole system, rather than just the compressor.

▶ 16.0 EVAPORATIVE COOLERS

16.1 INTRODUCTION

Function

Evaporative coolers are primitive air conditioners that push outdoor air into the home after adding water to the air to cool it. The unit may be mounted on an exterior wall or on the roof. There is usually a simple, supply-only duct system to distribute the cooled moist air through the home.

Dry Climates Only

As discussed in **The Home Reference Book**, these systems are only used in hot, dry climates. They rely on the latent heat of vaporization to lower the temperature of the air as the air evaporates water. These units can be thought of as large humidifiers. While increased humidity is not desirable in most air conditioning climates, if the air is very dry, lowering the air temperature by adding moisture is not a problem.

Swamp Coolers

Some people refer to these as **swamp coolers** because they do increase the humidity levels inside the house. Because of the standing water, there may be some health issues associated with algae, bacteria, etc., in the water.

Use Outdoor Air In Single Pass System

Air is drawn in from outside the building and passed through the evaporative cooler before being discharged into the building. The coolers are single-pass systems. The air is not recirculated because once it has picked up moisture, it can't be cooled much more by passing it through the cooler a second time. There is no return duct system. **Windows in the house often have to be open to allow air to exhaust from the house.**

The water is recirculated and the water level is topped up automatically as needed.

Components

All evaporative coolers have an electrically operated air handler, a water reservoir, an overflow, a water makeup valve and float, a pump and some way to get water into the air stream.

Three types of evaporative coolers are **rotary, spray** and **drip**.

1. Rotary cooler

The **rotary evaporative cooler** has a drum that rotates through a tank of water picking up water and allowing it to drip off, like a very fine paddle wheel. The drum is made of layers of metal screening that pick up the water from the reservoir and drops the water into the air stream.

SECTION ONE: AIR CONDITIONING

2. Spray cooler

The **spray type cooler** has a water spray directed up into the air stream. The spray may be absorbed by the air stream or collected on an evaporator pad. The water that accumulates on the pad may be evaporated into the air stream, or may run back down into the reservoir. Many coolers have a pad that is kept wet. The air moves through the pad.

3. Drip Cooler

The **drip type evaporative cooler** employs spider tubes and bleeders that allow the water to drip onto the pads.

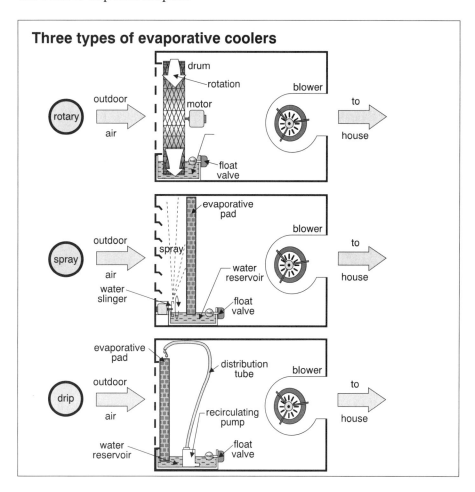

Water In Reservoir Controlled By Float Valve

Units typically have a **float valve** that allows the water level to be maintained in the reservoir at a certain level. In some jurisdictions, an **air gap** is required to prevent a cross connection and contaminated water from the reservoir flowing back into the potable water. Many units have a **reservoir overflow drain**.

SECTION ONE: AIR CONDITIONING

Bleed Off And Auto Cleaning
Some coolers have a **bleed-off** system which changes the water in the reservoir by continuously draining off some of the water. This can lead to high water consumption. The **auto cleaning system** is an alternative to the bleed-off. It pumps the water out of the reservoir into the overflow. The pump typically operates every 5 to 12 hours. The idea is to keep the reservoir water fresh and clean, but use less water than a bleed-off system.

Blower
The electric blowers may be one-speed or two-speed.

Wall Or Roof Mounted
The units may be mounted on walls or roofs. In either case, they have to be suitably flashed and weather-tight so that rain will not enter the building. Wall-mounted units should be at least 6 inches above grade.

The fan motors may be $1/3$ to $3/4$ horsepower and commonly move 3,000 to 6,000 CFM. This is much more than a 3-ton split-system central air conditioning unit that moves roughly 1,300 to 1,500 CFM. A maximum air velocity of about 500 linear feet per minute is desirable.

Water Consumption
Evaporative coolers consume a fair bit of water. To achieve a temperature drop of 10°F, an evaporative cooler might use $1¼$ gallons of water per hour for every 1,000 CFM blowing across the cooler.

Pads
The pads that catch the water and transfer it to the air are commonly made from shredded wood (aspen) or a proprietary system such as CEL dek® or GLAS dek®. CEL deck® is made from high cellulose paper and GLAS dek® is made from glass matts.

Drying Pads
Many recommend that pads should dry out once every 24 hours to control algae growth. Drying them too often (e.g., every hour) may promote scale accumulations.

The entire pad area should be wetted for best performance and longest pad life.

Winterized
In cold climates, evaporative coolers are shut down for the winter. The outdoor part should be covered and the water should be shut off and drained to prevent freezing and subsequent flooding.

Controls
The controls for an evaporative cooler often have a high and low setting. There is also usually a setting that turns on the water and fan (cooling) or just the fan (fan only).

Pump Only
Many controls also have a "pump only" position. This allows the pump to flush dirt out of the pads and fully saturate the pads before pulling air through them. It is normally used on start-up when the system has been idle for some time.

Delayed Action
When starting a cooler, it takes some time for cooling to be effective. It may take 30 minutes for the pads to be fully saturated and the full benefits of cooling to be felt.

Inspecting the evaporative cooler also involves inspection of the visible ducts and registers inside the house.

SECTION ONE: AIR CONDITIONING

16.2 CONDITIONS

Common evaporative cooler problems include the following:

1. Leaking
2. Pump or fan inoperative
3. Rust, mold and mildew
4. No air gap on water supply
5. No water
6. Poor support for pump and water system
7. Louvers obstructed
8. Missing or dirty air filter (systems without pads)
9. Cabinet too close to grade
10. Cabinet or ducts not weather-tight
11. Electrical problems
12. Duct problems
13. Excess noise or vibration
14. Clogged pads

16.2.1 LEAKING

Causes Water may leak as a result of –

- a stuck float switch
- a rusted, split or tilted water tray
- a leaking valve or distribution line
- an obstructed drain
- poor connections on the water distribution system
- poor connections at the water supply line

Implications Water leakage can cause rusting in the cabinet and may damage interior or exterior finishes below. Water can also run through the ducts and damage other parts of the home.

Strategy Check the interior of the unit for ponding water and concentrated rust or moisture stains. Also look at the areas around and below the unit for evidence of rust, mold, mildew, water stains and, of course, standing or dripping water.

16.2.2 PUMP OR FAN INOPERATIVE

Causes The pump or the fan may not work because of –

- electrical problems
- mechanical problems

Implications If either the pump or the fan is not working, there will not be any effective cooling.

Strategy When you start up the system, ensure that the pump is moving water and the fan is moving air.

16.2.3 RUST, MOLD AND MILDEW

Cause Because water is an integral part of an evaporative cooler, rust, mold and mildew can be expected wherever there is leakage or standing water. Algae and bacteria can build up in the reservoir, for example.

Homes with evaporative coolers may experience buildup of mold and mildew as well as bacteria in the house itself. The warm air is cooled and humidified as it enters the house. If the temperature drops and the air cools, the moisture in the air may condense. This can lead to mold and mildew forming on cool surfaces in the house and in the duct system. There can be health implications to this.

Implications Rust will eventually destroy the metallic parts of the system. Rust may also clog the plumbing components and attack the electric motor and its connections. Mold and mildew are, of course, health issues.

Strategy Open the cabinet and examine all of the accessible components for rust. Pay particular attention to the cabinet floor and any areas below the water tray.

16.2.4 NO AIR GAP ON WATER SUPPLY

Cause This is an installation error. There should be a gap between the point where water is discharged from the supply system and the water level in the tray. The tray water may be 2 to 2½ inches deep. The overflow pipe should prevent water from rising higher than this. The supply water should be at least 3 to 3½ inches above the bottom of the tray.

Implication If the water in the system can get back into the supply water, this is a cross connection. People may be poisoned by contamination of their potable water.

Strategy Ensure that there is at least a 1-inch gap between the water discharge and the water level in the tray.

16.2.5 NO WATER

Causes There may not be any water in the system because –

- the float switch is inoperative
- the distribution system is clogged
- the water has been shut off at the source

Implications The system will not cool effectively without water.

Strategy Ensure that there is water flowing when the system is operating and that there is water in the pan.

16.2.6 POOR SUPPORT FOR PUMP AND WATER SYSTEM

Cause This may be caused by –

- poor installation
- rust damaging the supports
- mechanical damage to the supports

Implications The system may fail to operate or flood as a result. There could also be an electrical hazard if electrical components get wet.

Strategy Check that the pump and the water distribution system are well secured.

16.2.7 LOUVERS OBSTRUCTED

The louvers need to allow outside air to be drawn in through the moistened pads and distributed throughout the house.

Causes Louvers may be obstructed by rust, mechanical damage or adjacent vegetation.

Implications There won't be effective cooling if the louvers do not allow good airflow.

Strategy Make sure that the louvers allow good airflow when the system is operating.

16.2.8 MISSING OR DIRTY AIR FILTER

Cause This is a maintenance issue.

Implication Poor airflow and inadequate cooling will result if the filter is dirty. The implication of a missing filter is poor air quality as exterior dirt and dust will be drawn in and blown throughout the home.

Strategy Make sure there is a clean filter in place.

16.2.9 CABINET TOO CLOSE TO GRADE

The cabinet should be at least 6 inches above grade.

Cause This is an installation or building settlement issue in most cases. In some cases it can be a landscaping problem.

Implication The cabinet is prone to rusting if it is too close to grade.

Strategy Make sure the cabinet is at least 6 inches above grade. Where the cabinet is near grade level, check for exterior rust around the base of the cabinet.

SECTION ONE: AIR CONDITIONING

16.2.10 CABINET OR DUCTS NOT WEATHER-TIGHT

Cause This is an installation or maintenance issue.

Implications Water leakage into the building is the implication.

Strategy Make sure the points of penetration through the wall or roof are weather-tight.

16.2.11 ELECTRICAL PROBLEMS

Many jurisdictions require an electrical disconnect outside the building within sight of and readily accessible to the cabinet. The electrical system to the cooler should be grounded.

Cause Electrical problems are usually installation issues.

Implications A convenient disconnect is a requirement for safe servicing of the unit. If it is not convenient to shut the unit off when working on it, people will be tempted to work on it with live electrical connections. There is a life safety hazard here.

A grounding system is important wherever water and electricity are together. Again, this is a life safety issue. It's better to have electricity go through a ground wire than through you.

Strategy Make sure there is a disconnect close to the cabinet.

16.2.12 DUCT PROBLEMS

Not Connected To Heating Ducts — For the most part, the duct problems are the same as what we would find on any air conditioning system. There is one issue particular to evaporative coolers: the cooler ducts cannot be interconnected with the heating ducts. There is a danger of rusting out a heat exchanger on a furnace.

Cause This is an installation issue.

Implications Premature failure of the furnace is the implication.

Strategy Try to follow the duct system and make sure the ducts for the evaporative cooler are independent of the heating ducts. By checking the airflow at the registers with different equipment running, you should be able to verify the separation of the duct systems.

SECTION ONE: AIR CONDITIONING

16.2.13 EXCESS NOISE OR VIBRATION

Causes This may be a pump or fan problem. It is usually caused by –

- bad bearings
- poor attachment
- poor alignment of motor and fan or motor and pump

Implications Premature failure of these components is likely. The noise may also be unacceptable to the occupant.

Strategy When the system is operating, listen for excess noise and watch for excess vibration.

16.2.14 CLOGGED PADS

The pad/media can be clogged with mineral buildup, dirt, algae, etc.

Cause This is caused by poor maintenance, dirty filters, and dirty water.

Implications Clogged pads won't do too much cooling.

Strategy Check the airflow through the pads and at the registers.

No Exhaust Openings Sometimes poor airflow isn't the result of dirty filters or pads. It may be that air in the house can't escape. Check that some windows are open so the warm air can be exhausted. As a general rule, there should be about two square feet of opening for every 1,000 CFM of cooler capacity.

Summary

Evaporative coolers are relatively simple and primitive from an air conditioning standpoint. They can however, have several problems and can create water damage to the home if not well maintained. Evaporative coolers are effectively large humidifiers.

▶ 17.0 WHOLE-HOUSE FAN

17.1 INTRODUCTION

Function Whole-house fans aren't really air conditioners but they do help to cool houses. Whole-house fans draw air in through the house windows and exhaust through an opening, often in the ceiling at the top level of the house. The air is discharged into the attic and is exhausted out through the roof, soffit and gable vents.

Not Air Conditioners These fans are typically quite powerful and, while they do not condition the air, moving the air through the house quickly enhances evaporative cooling across people's skin and also prevents the air in the house and attic from getting considerably warmer than the outdoor air. The attic temperatures can easily exceed the outdoor temperatures. This hot air trapped in the attic heats up the house as well. A whole-house fan minimizes attic heat buildup.

SECTION ONE: AIR CONDITIONING

Louvered Opening In Ceiling

The opening in the ceiling is usually louvered, with the fan often located just above it. As the fan starts to turn, the louvers are drawn open, exhausting air from within the house. When the fan shuts off, the louvers drop closed. In heating climates, this opening is leaky and considerable energy can be lost up into the attic. Some installations include an insulated box built around the fan. The box is placed over the fan in the winter months and removed during the summer months.

Manual Or Thermostatic Control

Whole-house fans can be operated manually through a wall switch, typically located on a upper floor hallway near the fan itself, or the fan can be operated thermostatically. For example, if the house (or attic) temperature is more than 75°F or 80°F, the fan can be set to come on.

Risk Of Backdrafting

As discussed in **The Home Reference Book**, there is the possibility of backdrafting combustion appliances with a whole-house fan. Care should be exercised in operating these fans to ensure that backdrafting of fireplaces, water heaters, etc., does not occur. Generally speaking, backdrafting is avoided by opening several windows in the house. Operating a whole-house fan with windows and doors closed will almost certainly backdraft any natural draft combustion appliance, as the fan puts the house under significant negative pressure.

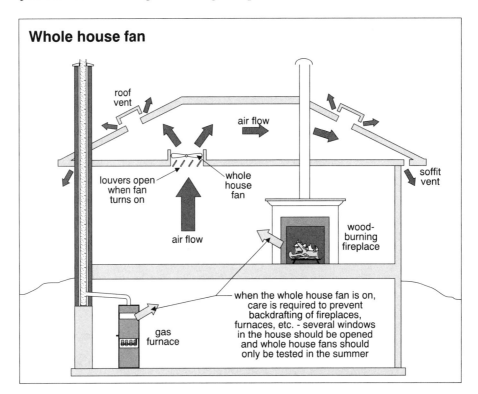

These are sometimes called **attic fans** and, in some cases, the air supply to the fan may be taken from ducts and grilles in several rooms, rather than a single ceiling grille.

The outlet area for an attic fan is ideally 1½ times the area of the fan intake. Under no circumstances should the outlet area be less than the fan intake area.

SECTION ONE: AIR CONDITIONING

17.2 CONDITIONS

Common whole-house fan problems include the following:

1. Inoperative, noisy or excess vibration
2. Inadequate attic venting

17.2.1 INOPERATIVE

If the fan does not start, there are several possibilities.

Causes
- The electrical source to the fan may have been interrupted.
- The fan motor may be burned out.
- The belt on belt-driven fans may be slipping or broken.
- The fan may be obstructed and unable to turn.
- The fan may be controlled by a manual switch plus a thermostat.
- The thermostat may not be calling for the fan to come on.

Excess noise or vibration may be due to –

- bad bearings
- fan/motor misalignment
- a slipping belt
- loose mounts
- dirty or bent fan blades

Implications The whole-house fan is not an essential component to the home. The house will be less comfortable in hot weather if the fan does not work.

Test In Summer Only Whole-house fans should not be tested in the winter. They should only be tested during warm months when a combustion heating system is not likely to be in service. Before operating the whole-house fan, check that an insulating cover is not in place. Also, make sure there are several windows or doors open before activating the fan.

If you have verified that it is safe to turn on the fan, flip the switch and the fan should start. Most of these fans are quite noisy, and you will see the louvers in the ceiling open.

Strategy If the fan does not respond by flipping a switch, look for a thermostat mounted either in the house or in the attic. Lowering the setting on the thermostat will activate the fan if it is working properly. Be sure to reset the thermostat to its original setting.

Listen for a hum that suggests the motor cannot overcome start-up inertia. If a hum is heard, the system needs servicing, which may include lubrication, cleaning, re-alignment, belt adjustment or motor replacement.

If it does not operate, further investigation should be recommended.

If the fan operates but is excessively noisy or vibrates too much, recommend servicing. It will take some experience to know what is typical.

SECTION ONE: AIR CONDITIONING

17.2.2 INADEQUATE ATTIC VENTING

In some installations, a whole-house fan has restricted discharge, owing to very little attic venting.

Causes This is usually a result of poor installation.

Implications The fan will not move as much air as it should, and will have to work harder, shortening the life expectancy of the equipment.

Strategy Check that there is at least as much attic ventilation as one would expect to find in a well insulated and ventilated ceiling. One square foot of venting for every 300 square feet of attic is an absolute minimum. This venting can be through the soffits, the gables or the roof itself.

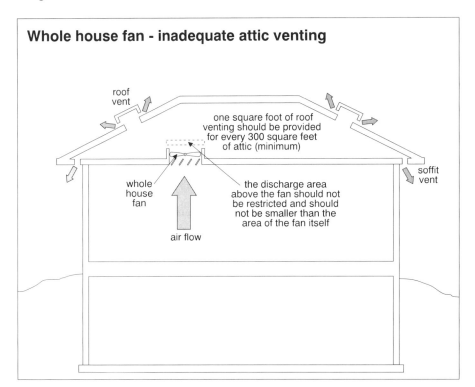

We said earlier that the total exhaust area should be at least 1½ times the fan intake.

If the whole-house fan operates, but the louvers do not open, or only open slightly, it may be because too much positive pressure is established in the attic. Check for adequate exhaust ventilation.

When you operate a whole-house fan, you can usually feel the air blowing by you as you stand near its intake point in the house. If you can't feel the air, there may be an obstruction on the exhaust side. Other possibilities include a fan belt that is slipping or otherwise working improperly.

SECTION ONE: AIR CONDITIONING
S5

Air Conditioning & Heat Pumps
MODULE

QUICK QUIZ 5

☑ INSTRUCTIONS

- You should finish Study Session 5 before doing this Quiz.
- Write your answers in the spaces provided.
- Check your answers against ours at the end of the Section.
- If you have trouble with the Quiz, reread the Study Session and try the Quiz again.
- If you did well, it's time to move on to the Inspection Tools List

1. Give some examples of good locations and poor locations for thermostats.

2. List seven common problems with thermostats.

SECTION ONE: AIR CONDITIONING

3. Why does it matter if a mechanical thermostat is not level?

4. How long do air conditioning compressors typically last in your area?

5. What is the typical warranty period?

6. When compressors in split systems fail, can they be replaced or does the whole air conditioning system have to be replaced?

7. What is another name for an evaporative cooler?

8. List three types of evaporative coolers.

9. Evaporative coolers recirculate the house air across the cooler.
   True ☐   False ☐

10. Explain why evaporative coolers need an air gap.

11. Where are whole-house fans typically located?

SECTION ONE: AIR CONDITIONING

12. Name two ways that they might be activated.

_____

_____

13. List two common whole-house fan problems.

_____

_____

If you didn't have any trouble with this Quiz then it is time to look at the Inspection Tools.

Key Words:
- **Central thermostat**
- **Compressor life**
- **Compressor warranty**
- **Evaporative coolers**
- **Swamp coolers**
- **Single pass**
- **Rotary, spray and drip**
- **Rust**
- **Air gap**
- **Whole-house fan**
- **Louvers**
- **Backdrafting**
- **Combustion appliances**
- **Attic venting**

▶ 18.0 INSPECTION TOOLS

Flashlight
A good light is needed to look at the air conditioning system, particularly the evaporator coil and fan. A flashlight can also be helpful when looking inside the condenser unit at the condenser coil and compressor. The brighter the light, the better. Some inspectors use trouble lights (incandescent lights on cords) to inspect air conditioners.

Telescopic mirror
Mirrors are helpful when looking inside cabinets and plenums. Mirrors are almost always used with a flashlight.

Thermometer (with 6- inch probe-type sensor)
This can be used to check the temperature drop across the evaporator coil. This goes beyond the Standards but is a common test among home inspectors.

Screwdrivers and pliers
Although you shouldn't have to use tools to remove most covers, these tools are sometimes helpful to persuade stiff covers. They are also useful if you choose to go beyond the Standards and remove condenser cabinet covers, for example.

Rags or towels
Air conditioners are often dirty. You should carry something to clean your hand when you are finished your inspection of the air conditioner. Some inspectors carry wet wipes. Most inspectors avoid using the washroom in the house for clean up.

Tape measure
A tape measure can be useful when approximating the house square footage. However, most inspectors step-off these distances. You can also use the tape to check supply and return plenum sizes.

Clamp-on ampmeter
The ampmeter is used to measure the current draw of the compressor and compare it to the data plate. This is another test that goes beyond the Standards.

Your senses
As always, your eyes are your main inspection tool. You'll be using your sense of touch to feel airflow and temperature at registers and at the outdoor fan, for example. You'll also be touching Freon lines to feel temperatures there. Even without trying, you body will feel either warm or cool as you do you inspection inside the house. This will give you a general sense of the performance of the heating and cooling functions.

Tissues
Since the airflow through return registers is often too low a volume to sense with the hand, many inspectors use tissue to check for airflow through the return grilles.

SECTION ONE: AIR CONDITIONING

▶ 19.0 INSPECTION CHECKLIST

Location Legend   N = North   S = South   E = East   W = West
1 = 1st Floor   2 = 2nd Floor   3 = 3rd Floor   B = Basement   CS = Crawlspace

	CENTRAL AIR CONDITIONING		
LOCATION	**CAPACITY**	**LOCATION**	**REFRIGERANT LINES**
	• Undersized		• Leaking
	• Oversized		• Damaged
			• Missing insulation
	COMPRESSOR		• Lines too warm or too cold
	• Excess noise/vibration		• Lines touching each other
	• Short cycling		• Low points or improper slope in lines
	• Running continuously		**EXPANSION DEVICE**
	• Out of level		• Capillary tube crimped/ disconnected/leaking
	• Excess electric current draw		• Thermostatic expansion valve loose/clogged/sticking
	• Wrong fuse or breaker size		
	• Electric wires too small		**CONDENSER FAN**
	• Missing electrical shut-off		• Excess noise/vibration
	• Inoperative		• Inoperative
	• Inadequate cooling		• Corrosion
	AIR COOLED CONDENSER COIL		• Mechanical damage
	• Dirty		• Obstructed air flow
	• Damaged		• Dirty
	• Corrosion		**EVAPORATOR FAN**
	• Clothes dryer or water heater exhaust too close		
			• Undersized
	WATER COOLED CONDENSER COIL		• Misadjustment of belt or pulleys
	• Leakage		• Excess noise/vibration
	• Cooled by pool water		• Dirty
	• No backflow preventer		• Dirty or missing filter
	• No water		• Inoperative
			• Corrosion
			• Damage
	EVAPORATOR COIL		**DUCT SYSTEM**
			• Missing
	• Temperature split too high		• Undersized
			• Incomplete
	• Temperature split too low		• Leaky or disconnected
			• Dirty
	• No access to coil		• Disconnected or leaking
	• Dirty		• Obstructed, damaged or collapsed
	• Frost		• Poor support
	• Top of coil dry		• Poor balancing
	• Corrosion		• Humidifier damper missing
	• Damage		• Supply/return registers — too few
			• Supply/return registers — poor location
	CONDENSATE SYSTEM		• Supply/return registers — obstructed
	• Pan leaking or overflowing		
	• Dirt in pan		**REGISTERS AND GRILLES**
	• Pan cracked		• Missing, stiff or inoperable
	• Inappropriate pan slope		• Dirty
	• Rust or holes in pan		• Obstructed or damaged
	• Pan not well secured		• Rust
	• No auxiliary pan		• Poor location
	• No float switch		• Not well secured
			• Weak airflow

SECTION ONE: AIR CONDITIONING

CENTRAL AIR CONDITIONING			
LOCATION	**DUCT INSULATION**	**LOCATION**	**VAPOR BARRIER**
	• Missing		• Missing
	• Incomplete		• Damaged
	• Damaged		• Incomplete
	CONDENSATE DRAIN LINE		**THERMOSTATS**
	• Leaking/damaged/split		• Inoperative
	• Disconnected/missing		• Poor location
	• Blocked or crimped		• Not level
	• No trap		• Loose
	• Improper discharge point		• Dirty
			• Damaged
	CONDENSATE PUMP		• Poor adjustment/calibration
	• Inoperative		
	• Leaking		
	• Poor wiring		

EVAPORATIVE COOLERS			
LOCATION	**DUCT PROBLEMS**	**LOCATION**	**GENERAL**
	• Leaking reservoir		• Missing/dirty air filter
	• Pump or fan inoperative		• Cabinet too close to grade
	• Rust, mold and mildew		• Cabinet/ducts not weather-tight
	• No air gap on water supply		• Electrical problems
	• No water		• Duct problems
	• Poor support for pump and water system		• Excess noise/vibration
	• Louvers obstructed		• Clogged pads
	• Overflow problems		• Drum or spray system problems

WHOLE-HOUSE FAN			
LOCATION	**GENERAL**	**LOCATION**	
	• Inoperative		
	• Excess noise/vibration		
	• Inadequate attic venting		

SECTION ONE: AIR CONDITIONING

▶ 20.0 INSPECTION PROCEDURES

20.1 AIR-COOLED AIR CONDITIONING

Our inspection procedure includes both a visual inspection and an operating test. We'll look at the visual inspection process first, and then move on to the operating tests for the various systems.

A. VISUAL INSPECTION

Outdoors

1. Find the outdoor unit and check the condenser cabinet for rust or damage.
2. Make sure that the condenser is level within ten degrees.
3. Make sure the unit is not partially buried in the soil.
4. Visually inspect the condenser coils for dirt, lint and other obstructions and for corrosion or fin damage.
5. Look for a clothes dryer or water heater vent within 6 feet of the unit.
6. What condition is the condenser fan in?
7. Is the airflow obstructed? (You should have 1 to 3 feet on the intake side and 4 to 6 feet on the discharge.)
8. Is there evidence or corrosion or damage?
9. Record the unit size, age, compressor Rated Load Amperage (RLA), maximum fuse size and minimum circuit ampacity from the data plate.
10. Check the unit size against the square footage of the house (general rule).
11. Look for an outdoor electrical disconnect within sight of and readily accessible to the condenser.
12. Check the size of the fuses or the breakers.
13. Check the circuit ampacity against the wire size. Check the condition of wires and connections. Check that outdoor wires are suitable for outdoor use.
14. For those who go beyond the Standards –
    - shut off the power
    - open the cabinet
    - record the age of the compressor
    - look for a bulging capacitor
    - look for oil in the cabinet indicating refrigerant leakage
    - look for dirt, fin damage and corrosion
    - check the electrical connections
    - close the cabinet unless you are going to check the electrical current draw when the system is operating
    - turn the power back on.

SECTION ONE: AIR CONDITIONING

Indoors and out

1. Check the refrigerant lines for insulation, oil deposits (indicating leakage) and evidence of crimping or mechanical damage. Refrigerant lines should not touch each other.

Indoors

1. Check for service access to the plenum coil. If it's available, open the access.
2. Check the size of the coil, if visible, and compare it to the condenser size.
3. Check the indoor coil for dirt, damage, corrosion or frost. Pay close attention to the upstream side. If the unit has been operating, make sure that the entire coil is wet.
4. Check the expansion device (capillary tube or thermostatic expansion valve) for evidence of leakage or mechanical damage.
5. Check the condensate (and auxiliary) drain pan for evidence of leakage, dirt, rust, improper slope or poor attachment. If there is no auxiliary pan, and the coil is above finished living space, a high water-level cutout (float switch) should be present.
6. Check the condensate line for a trap, evidence of leakage, blockage and appropriate discharge point. If there is a pump, check the wiring to the pump. Is pump grounded? Is there any evidence of leakage around the pump?

Note: Turn the power off and remove the blower cover.

7. Check the blower size against the furnace data plate (if applicable). Check the fan blades for dirt.
8. Check for a loose or worn belt or misalignment of the belt and pulleys.
9. Check for the presence and condition of an air filter.

Note: Replace the blower cover and turn the power back on.

Duct System

10. Are there vibration collars? (It's not a deficiency if they are missing, but a bonus if they are there.)
11. Is the duct system complete? (Are all parts of the home served?)
12. Are the ducts, registers and grilles clean?
13. Are any of the ducts disconnected or leaking?
14. Can you see any obstructed or collapsed ducts?
15. Are the ducts well supported?
16. Is there a humidifier on the system? If so, is there a damper on the humidifier duct? If so, is it in the correct position for the season?
17. Are ducts passing through unconditioned spaces insulated? Is the vapor barrier intact?
18. Are the supply and return grilles located in a suitable position for air conditioning? (Are they better suited to heating?)

SECTION ONE: AIR CONDITIONING

Thermostat
19. Is it well located?
20. Is it level?
21. Is it clean?
22. Is it well secured?
23. Is there any evidence of damage?
24. Is the temperature in the house close to the temperature that the system is set at?

B. OPERATING TEST

Note: If the furnace has been running, wait at least ten minutes after the house air fan stops before testing the air conditioning. This allows the refrigerant pressure to equalize through the system and avoids damaging the compressor.

1. Determine if the system has had power for at least 12 hours.
2. If the outdoor temperature is above 65°F (and has been for at least 12 hours or so), turn the thermostat down to its lowest setting. Listen for the house air fan coming on.
3. Check the condenser fan for unusual noise and vibration.
4. Check the compressor for unusual noise and vibration. Watch for short cycling.
5. After the system has been running for 15 minutes, check the air coming off the condenser coil with your hand. It should be warmer than the ambient air.
6. Check the temperature of the refrigerant lines.
7. If you go beyond the Standards and use an amp-meter, measure the electric current flow through the compressor. Compare it to the compressor Rated Load Amperage (RLA). It should be roughly 60 to 90 percent of the RLA. Replace cabinet cover.
8. Check the temperature difference (split) across the evaporator coil.
9. Check the evaporator coil for frost.
10. Check the condensate line to ensure water is flowing out. Look for leaks in the condensate pan and line.
11. If there is a float switch for the condensate pan, hold the float up so the unit thinks that the pan is full. The unit should shut off. This may take about five minutes. Release the float. The unit should re-start. This may take five to seven minutes.
12. If there is a condensate pump, check that it is working properly.
13. Replace or close the access cover for the coil.

14. Check the house air fan for unusual noise and vibration.
15. Check the ducts for leakage, missing insulation and compromised vapor barriers. Are ducts in attics and crawlspaces sweating?
16. Check the airflow at supply and return registers throughout the house.
17. Return the thermostat to its original position.

20.2 WATER-COOLED AIR CONDITIONING

A. VISUAL INSPECTION

Inspection procedures are the same as air-cooled systems, with some minor exceptions. The condensing coil and compressor will be inside the house. There is no condenser fan, because the condenser transfers heat from the refrigerant to water, rather than air.

1. Check the condenser for rust or evidence of leakage.
2. Record the unit size, age, compressor RLA, maximum fuse size and circuit ampacity.
3. Check unit size against square footage of house (general rule).
4. Locate the electrical disconnect. Check the size of fuses or breakers.
5. Check the circuit ampacity against the wire size. Check the condition of wires and connections.
6. Trace the supply and waste water piping. Look for multiple discharge points for the waste water (sewer, lawn, sprinkler system, swimming pool, other).
7. Check for a backflow preventer on the supply water line.
8. Check for swimming pool water coming to the condenser coil.
9. Ensure that any manual water supply valve is open.
10. Check the refrigerant lines for insulation, oil deposits (indicating leakage) and evidence of crimping or mechanical damage. Refrigerant lines should not touch each other.
11. Check for service access to the plenum coil. If it's available, open the access.
12. Check the size of the coil, if visible, and compare it to the condenser size.
13. Check the indoor coil for dirt, damage, corrosion or frost. Pay close attention to the upstream side. If the unit has been operating, make sure that the entire coil is wet.
14. Check the expansion device (capillary tube or thermostatic expansion valve) for evidence of leakage or mechanical damage.
15. Check the condensate (and auxiliary) drain pan for evidence of leakage, dirt, rust, improper slope or poor attachment. If there is no auxiliary pan and finished areas may be damaged by leakage, a high water-level cutout (float switch) should be present.

SECTION ONE: AIR CONDITIONING

16. Check the condensate line for a trap, evidence of leakage, blockage, and appropriate discharge point. If there is a pump, check the wiring to the pump. Is the pump grounded? Is there any evidence of leakage around the pump?
17. Check the blower size against the furnace data plate (if applicable). Check the fan blades for dirt.
18. Check for a loose or worn belt or misalignment of the belt and pulleys.
19. Check for the presence and condition of an air filter.

B. OPERATING TEST

Note: If the furnace has been running, wait at least ten minutes after the house air stops before testing the air conditioning. This allow the refrigerant pressure to equalize through the system and avoids damaging the compressor.

1. Turn the thermostat to its lowest setting.
2. Listen for the house air fan and compressor to come on.
3. Check for waste water flow at the discharge point.
4. The temperature of the waste water line should be higher than the supply water line.
5. Check the compressor for unusual noise and vibration.
6. Check for water leaks.
7. Check the temperature of the refrigerant lines.
8. If you go beyond the Standards and use an ampmeter, measure the electric current flow through the compressor. Compare it to the RLA.
9. Check the temperature difference (split) across the evaporator coil.
10. Check the evaporator coil for frost.
11. Check the condensate line to ensure water is flowing out. Look for leaks in the condensate pan and line.
12. Lift up the float on the high water cut-out switch to shut-off the unit. This may take five minutes. Release the float and the unit should come back on in five to seven minutes.
13. If there is a condensate pump, check that it is working properly.
14. Replace or close the access cover for the coil.
15. Check the house air fan for unusual noise and vibration.
16. Check the ducts for leakage, missing insulation and compromised vapor barriers. Are ducts in attics and crawlspaces sweating?
17. Check the airflow at supply and return registers throughout the house.
18. Do your flow test of the house plumbing fixtures while the unit is running. (Water pressure/flow may be low.)
19. Return the thermostat to its original setting.

SECTION ONE: AIR CONDITIONING

20.3 EVAPORATIVE COOLERS

A. VISUAL INSPECTION

Outdoors
1. Check the cabinet for rust or damage. Ensure that the louvers are not obstructed.
2. Ensure the unit is at least 6 inches above grade level.
3. Check for weather-tightness at the junction of the cabinet or ductwork and the building skin. Check the condition of mounting system to roof.
4. Look for an outdoor electrical disconnect accessible to and in sight of the unit.

Note: Shut off the power.

5. Open the cabinet.
6. Check the inside of the cabinet for rust, moisture, water stains, mold and mildew.
7. Check the air handler for rust, dirt on the fan blades, a loose or worn belt, misalignment of belts and pulleys, and electrical connections.
8. Check the water reservoir for leakage or obstructions. Is there water in the pan? Is it clean?
9. Check the pump and water distribution system for condition and support.
10. Check the water supply line, overflow and float and ensure that a 1-inch air gap is present.
11. Check the pads. Are they clogged or dried out?
12. Check the air filter.

Note: Turn the power back on and close the cabinet.

Indoors
1. Check the ducts and registers for rust, obstructions, damage and interconnection with heating ducts.
2. Locate the operating controls.

B. OPERATING TEST

1. Activate the unit in each mode.
2. Check for proper water flow and watch for leakage.
3. Is the fan one- or two-speed?
4. Check the airflow through the registers.
5. Listen and watch for unusual fan noise or vibration.
6. Turn the system off.

20.4 WHOLE-HOUSE FAN

A. VISUAL INSPECTION

1. Find the whole-house fan and the operating controls.
2. Are there adequate discharge opportunities for the air after it is pulled out of the house?
3. Is the attic well ventilated?
4. Are there combustion appliances that might be affected by turning the whole-house fan on? If so, then do not operate the fan. (Operating these systems in the winter is risky. You may backdraft a furnace or boiler burner.)

B. OPERATING TEST

5. Does the unit have an insulating cover on it to prevent heat loss during the winter? If so, do not operate the unit.
6. Do not operate the unit unless there are enough windows open to draw fresh air into the home. This will help to avoid putting the house under excess negative pressure.
7. When the system starts up, do the louvers open freely?
8. Does the system respond properly to its controls?
9. Is there excess noise or vibration?
10. When the system shuts off, do the louvers fall back in to place and close properly?

SECTION ONE: AIR CONDITIONING

Air Conditioning & Heat Pumps
MODULE

FIELD EXERCISE 1

☑ INSTRUCTIONS

For this Field Exercise, find some houses with central air conditioning. Depending or where you live, that may be a challenge. If so, it should give you a sense of how much emphasis you should place on this section.

Ideally, you'll find some houses with air-cooled and some with water-cooled air conditioners.

If you are in an area where evaporative coolers are used, you should find at least two houses with these.

You should also find at least two houses with whole-house fans, if possible. Again, these may or may not be common in your area.

You should allow yourself at least three hours for this Field Exercise. This exercise should be done when outdoor temperatures are above 65°F. We can't operate most central air conditioners when the temperature is lower than that. Approach this as you have several of the other Field Exercises:

- Identify the system.
- Look at the individual components.
- Look for specific conditions.
- Operate the systems.
- Speak to experts in the field.

Exercise A — Central Air Conditioners

In each house with central air conditioning, follow the Inspection Procedure outlined in Section 20. Use the appropriate section (air-cooled and water-cooled).

SECTION ONE: AIR CONDITIONING

Exercise B - Evaporative Coolers

Here we are going to be looking at a unit which is typically located outside on a wall or on the roof. We also have a rudimentary duct system inside.

Follow the Inspection Procedure set out in Section 20.

Exercise C - Whole House Fans

Follow the Inspection Procedures for whole-house fans set out in Section 20.

Exercise D - Writing a report

Select the central air conditioning system where you found the most difficulties. Write a short narrative report to your client that describes –

- the system you observed
- any conditions that will affect performance
- any limitations to your inspection, and
- an action list for the clients.

Exercise E

Speak to local air conditioning contractors and supply houses. Ask about the most common systems in your area.

1. Are there any systems that are known to be problematic?
2. What are the most common types of distribution systems?
3. Are water-cooled units common?
4. Are evaporative coolers common?
5. Are whole-house fans popular?
6. Are gas chillers ever found in your area?
7. What is the normal sizing rule in your area? (How many square feet per ton of air conditioning are usually cooled?)
8. What insulation levels are recommended on ductwork passing through unconditioned areas?
9. Are there many oversized air conditioners in your area?
10. Are condensers usually set on the ground or hung on building walls?
11. What is the most popular type of compressor used?
12. Do most compressors have sump or crackcase heaters?
13. How long do compressors typically last in your area?
14. Is there a best or worst manufacturer of compressors?

SECTION ONE: AIR CONDITIONING

15. What kind of clearances are normally required for the inlet and outlet on a condenser coil?

16. How far away does a clothes dryer vent have to be from a condenser coil?

17. If water-cooled units are common, where is the water usually dumped? Is pool water ever recirculated through condenser coils?

18. What are the most common expansion devices?

19. Is it common for residential systems to have sight glasses and filter/dryers?

20. How do you determine if refrigerant is leaking without the use of tools, test devices or gauges?

21. If the evaporator coil is higher than the condensing unit, are there any special considerations for running the refrigerant lines?

22. If there is excess length in the refrigerant lines, does it matter if they are coiled horizontally or vertically?

23. Are refrigerant lines sealed where they pass through the building wall?

24. Are most evaporator coils provided with a readily removable access cover?

25. How common a problem are clogged evaporator coils?

26. Are condensate drain lines typically trapped? Why or why not?

27. Are condensate pumps common?

28. Are auxiliary condensate pans common?

29. Are float switches a common alternative to auxiliary drain pans?

30. Where do main and auxiliary condensate lines usually discharge?

31. Are ducts sized differently for cooling and for heating?

32. What is the most common supply register and return grille location in rooms with air conditioning?

33. Is ductwork buried in concrete slabs common?

34. Are there problems with it?

35. Are there a lot of independent air conditioning systems in this area (Space Pak, etc)?

When you have completed the Field Exercise, you are ready for the Final Test.

SECTION ONE: AIR CONDITIONING

▶ 21.0 BIBLIOGRAPHY

TITLE	AUTHOR	PUBLISHER
Home Heating & Air Conditioning Systems	James L. Kittle	McGraw Hill
Heating, Ventilating and Air Conditioning Library, Volumes I, II & III	James E. Brumbaugh	Macmillan
Various papers	Compilation	Heating, Refrigerating and Air Conditioning Institute of Canada
1997 ASHRAE® Handbook Fundamentals	Compilation	American Society of Heating, Refrigerating and Air-Conditioning Engineers, Inc.
1996 HVAC Systems and Equipment	Compilation	American Society of Heating, Refrigerating and Air-Conditioning Engineers, Inc.
1995 HVAC Applications	Compilation	American Society of Heating, Refrigerating and Air-Conditioning Engineers, Inc.
Advanced Air Conditioning — Beyond the Basics (a paper)	Mark Cramer	Mark Cramer
Heat Pump Manual	Compilation	The Electric Power Research Institute
Heat Pump Systems	H.J. Souer and R.J. Howell	John Wiley and Sons
Heat Pumps: An Efficient Heating and Cooling Alternative	D. McGuigan	Garden Way Publishing Company
Heat Pump Primer	Compilation	Ontario Hydro

SECTION ONE: AIR CONDITIONING

▶ ANSWERS TO QUICK QUIZZES

Answers to Quick Quiz 1

1. a. and b.
2. False
3. False
4. True
5. Electricity – 240 volts
6. True
7. b.
8. Gas, low temperature and pressure
9. 50° F
10. 20 to 40° F
11. 50° F
12. 170 to 230° F
13. 170 to 230° F
14. 100° F
15. 100° F
16. 20 to 40° F
17. Gas
18. Liquid
19. False
20. False
21. Suction line
22. Sensible heat is sensed by a thermostat; when the temperature changes, sensible heat changes.
23. Latent heat occurs when heat is added or taken away without a change in temperature, such as when boiling water changes to steam.
24. 970 BTUS
25. Evaporative cooling removes moisture from people's bodies, which also removes heat. As moisture evaporates, the air immediately adjacent to the skin is saturated. Blowing the saturated air away allows more moisture to evaporate from our skin, making us feel more comfortable.

SECTION ONE: AIR CONDITIONING

26. As the warm air passes over the evaporator coil, it cools and loses its ability to carry moisture. That moisture is collected on the coil, making the cooler air dryer.

27. Condensate is collected in a pan under the evaporator coil, and drained to a tube that runs to the floor drain, for example.

28. It runs from the compressor discharge through the condenser to the expansion device inlet.

29. It runs from the expansion device outlet through the evaporator to the compressor inlet.

30. Liquid line

31. Suction line

Answers for Quick Quiz 2

1. 12,000 BTU/hr

2.
   1. Outdoor temperature
   2. Outdoor humidity
   3. Insulation level in the house
   4. Single, double, triple pane windows
   5. Whether windows are low E.
   6. Whether window coverings are open
   7. Amount of shading from trees
   8. Roof overhang
   9. Awnings or buildings nearby
   10. Amount heat generated inside by people and equipment
   11. Amount of east and west facing glass.

3. 450 to 700 square feet

4. 700 to 1000 square feet

5. False

6. True

7. True

8. 15-20°F

9. True

10. False

11. It keeps the oil at the base of the compressor warm enough to boil off the refrigerant.

SECTION ONE: AIR CONDITIONING

12. 1. Excessive noise/vibration
    2. Short cycling or running continuously
    3. Out of level
    4. Excess electric current draw
    5. Wrong fuse or breaker size
    6. Electric wires too small
    7. Missing electrical shut off
    8. Inoperative
    9. Inadequate cooling

Answers for Quick Quiz 3

1. Outdoors

2. Indoors (above the furnace)

3. Condenser fan is located outdoors

4. Receivers collect condensed refrigerant from the condenser coil.

5. 1. Dirty
   2. Damaged or leaking
   3. Corrosion
   4. Dryer or water heater exhaust too close

6. An air-to-air condenser transfers its heat to the outside air with the help of a fan. A water-cooled condenser transfers the heat to water.

7. 1. Leakage
   2. Coil cooled by pool water
   3. No backflow preventer
   4. Low plumbing water pressure

8. A-coil, slab coil, vertical coil

9. False

10. 1. No access to coil
    2. Dirty
    3. Frost
    4. Top of evaporator dry
    5. Corrosion
    6. Damage

11. The capillary tube acts as an obstruction in the line, reducing the pressure and temperature of the liquid Freon.

12. A TXV, or TEV acts as a more precise expansion device on larger air conditioners, or heat pumps.

SECTION ONE: AIR CONDITIONING

13. 1. Capillary tube defects
2. Thermostatic expansion valve connections loose
3. Clogged orifice
4. Expansion valve sticking

14. 1. The pan may be cracked, or have an open seam in it.
2. The opening to the drain line may be obstructed.
3. The drain line may be missing or disconnected.
4. The pan may not be sloped properly, allowing water to overflow rather than run to the drain.
5. The pan may be rusted through.
6. The pan may not be properly located below the coil.
7. The pan may be filled with debris, restricting the amount of water it can hold.

15. An auxiliary pan is needed when the evaporator coil is located in an attic, or anywhere over a finished living space.

16. Plastic or copper

17. The condensate line from the primary pan may or may not have a trap. Lines from secondary pans should never be trapped.

18. 1. The trap may be required if the condensate pan is upstream of the house air fan, because the water is under negative pressure.
2. If the condensate is allowed directly into waste plumbing, a trap is required.

19. Acceptable:
1. Outside the building directly into the ground
2. Near the basement floor drain (must be an air gap)

Unacceptable:
1. Into the granular fill beneath the basement floor slab
2. Into the waste plumbing stack

20. 1. Inoperative
2. Leaking
3. Poor wiring

21. Copper

22. False

23. False

24. True

SECTION ONE: AIR CONDITIONING

25. 1. Leaking
2. Damage
3. Missing insulation
4. Lines too warm or cold
5. Lines touching each other

26. This may be identified by an oil residue below the line set, typically at valves or coil connections.

27. Suction line (large insulated one)

28. Liquid line (small uninsulated one)

29. Suction line

Answers for Quick Quiz 4

1. The condenser fan is outdoors in the condenser unit.

2. The inlet temperature is outdoor air temperature, and outlet temperature is roughly 15 to 20 degrees F higher than the outdoor temperature

3. 1. Excessive noise/vibration
2. Inoperative
3. Corrosion or mechanical damage
4. Obstructed air flow

4. Indoors

5. It is also the furnace fan

6. Belt drive and direct drive

7. False

8. 1. Undersized blower or motor
2. Misadjustment of belt or pulley
3. Excessive noise/vibration
4. Dirty or missing filter
5. Inoperative
6. Corrosion/damage
7. Dirty fan

9. Poor house comfort, high energy costs, premature failure of heating and air conditioning components, dirt in the ducts and throughout the house.

10. False

SECTION ONE: AIR CONDITIONING

11. 1. Undersized or incomplete
2. Supply and return register problems
3. Dirty
4. Disconnected and leaking
5. Obstructed or collapsed
6. Poor support
7. Poor balance
8. Humidifier damper missing

12. False

13. False

14. Insulated, with appropriate vapor barriers.

15. The room will overheat because the bedroom becomes pressurized, preventing more cool air from entering.

16. There may be round plugs (1½ inches in diameter) in the ductwork, and 6-inch square metal patches near the furnace or air handler.

17. It shuts off the airflow through the humidifier in air conditioning season.

18. If it is left open, cool air could be drawn back into the cold air return and back over the evaporator coil, causing it to ice up, and reducing its efficiency.

19. Left open in the summer

20. 1½ to 2 inches typically, providing roughly R 7.

Answers for Quick Quiz 5

1. 1. Good location is near the center of the house, away from direct sunlight, fireplaces, light bulbs, etc.
2. Poor locations are in kitchens, near ductwork, ovens, etc., against chimneys, in direct sunlight, above televisions, or near doors and windows.

2. 1. Inoperative
2. Poor location
3. Not level
4. Loose
5. Dirty
6. Damaged
7. Poor adjustment/calibration

3. The set temperature of the thermostat will not be an accurate reflection of the room temperature, since the mercury bulb will not be level.

4. 1. 8 to 10 years – southern U.S.A.
2. 10 to 15 years – moderate climates
3. 15 to 20 years – northern climates

5. Five years typically

SECTION ONE: AIR CONDITIONING

6. Can replace the compressor only

7. Swamp cooler

8. Rotary, spray, and drip

9. False

10. To prevent a cross connection, and prevent the reservoir water flowing back into the potable supply.

11. In the ceiling on the top floor or in the attic

12. With a wall switch, or a thermostat

13. 1. Inoperative, noisy, or excess vibration
2. Inadequate attic venting

2 HEAT PUMPS

Air Conditioning & Heat Pumps

MODULE

SECTION TWO: HEAT PUMPS

► TABLE OF CONTENTS

	1.0	OBJECTIVES	2
STUDY SESSION 1	2.0	INTRODUCTION	5
	3.0	**HEAT PUMPS IN THEORY**	5
	3.1	The Concept	5
	3.2	The Mechanics	6
	3.3	Co-Efficient of performance (COP)	10
	3.4	When the heat pump cannot keep up	12
	3.5	Humidity is the enemy	14
	3.6	Cost effectiveness	15
	3.7	One little twist	15
STUDY SESSION 2	4.0	**HEAT PUMPS IN PRACTICE**	22
	4.1	Undersized or oversized	22
	4.2	Heat pumps are similar to air conditioners	23
	4.3	Differences between air conditioners and heat pumps	23
	4.4	Heat pumps differ from gas and oil furnaces	26
	4.5	Defrost cycle	26
	4.6	Identifying heat pumps	28
	4.7	Conditions	29
	4.7.1	Oversized for cooling and undersized for heating	33
	4.7.2	Inoperative in heating or cooling mode	33
	4.7.3	Poor outdoor coil location	34
	4.7.4	Coil iced up	35
	4.7.5	Airflow problems in house	36
	4.7.6	Back-up heat problems	37
	4.7.7	Old	39
	4.8	Other types of heat pumps	39
	4.8.1	Water-to-air system	39
	4.8.2	Earth-to-air system	40
	4.8.3	Solar systems	42
	4.8.4	Bivalent systems	43
	5.0	INSPECTION TOOLS	51
	6.0	INSPECTION CHECKLIST	52
	7.0	INSPECTION PROCEDURE	54
FIELD EXERCISE 1			
	8.0	BIBLIOGRAPHY	64
		ANSWERS TO QUICK QUIZZES	65

SECTION TWO: HEAT PUMPS

▶ 1.0 OBJECTIVES

In this section, we look at heat pumps. This is an extension of our discussion of central air conditioning. Heat pumps share many of the same components with air conditioners. Heat pumps, in fact, act as air conditioners during the summer months.

We are going to look at the –

- principle of operation of heat pumps
- co-efficient of Performance (COP)
- the balance point
- auxiliary heat

By the end of this section, you should be able to –

- identify heat pumps
- recognize all of the important heat pump components and be able to describe their functions in one sentence
- list the common problems found with heat pumps in general and with each of the components
- describe in one sentence the implication of nonperformance of any component
- list the common causes of failure or nonperformance for each component
- describe the inspection strategy and tools necessary to identify common problems with each component

Not The Last Word As always, our depth is sufficient to allow the general practitioner home inspector to perform a visual inspection and an operating test of a heat pump. We are not shooting for service technician level and there is always more material that you can study and courses that you can take. We also believe that it is a good idea to speak to the specialists in your area.

SECTION TWO: HEAT PUMPS

Air Conditioning & Heat Pumps
MODULE

STUDY SESSION 1

1. The first Study Session outlines the concept and mechanics of heat pumps. We'll also talk about the coefficient of performance (COP) and the balance point. We'll talk about auxiliary heat and how humidity is the enemy. We'll also touch on the cost effectiveness of heat pumps and some of the variations on the basic system.

2. At the end of this Session, you should be able to –

    - Explain in three sentences how the heat pump works in the cooling mode
    - Explain in three sentences how the heat pump works in the heating mode
    - Sketch the heat pump loop, locating the main components correctly
    - Label the sketch with the relative temperatures and pressures of the refrigerant at various points in the loop
    - Describe in one sentence how heat pumps are sized
    - Define **coefficient of performance** in one sentence
    - Define **balance point** in one sentence
    - Describe two common types of back-up heat
    - Explain in two sentences the need for defrosting
    - Name two types of heat pumps other than air-to-air

4. This Study Session may take you roughly 60 minutes.

5. Quick Quiz 1 is included at the end of this Session. Answers may be written in your book.

SECTION TWO: HEAT PUMPS

Key Words:
- *Air conditioning in reverse*
- *Outside coil*
- *Indoor coil*
- *Expansion device*
- *Compressor*
- *Refrigerant*
- *Latent heat of vaporization*
- *Reversing valve*
- *Condenser coil*
- *Evaporator coil*
- *Coefficient of performance*
- *Balance point*
- *Back-up heat*
- *Frost*
- *Water-source heat pump*
- *Ground-source heat pump*

SECTION TWO: HEAT PUMPS

▶ 2.0 INTRODUCTION

Complex Inspecting central air conditioning is hard enough. Inspecting heat pumps is even more challenging. Not only is it an air conditioning system that sometimes acts as a heating system, there is often a back-up heating system to inspect as well. Doesn't seem fair, does it?

Well, we've got to do it, so we'll start with the basics again and then move through the system components and their problems. We'll finish with an inspection procedure and checklist.

▶ 3.0 HEAT PUMPS IN THEORY

3.1 THE CONCEPT

In many parts of North America, air conditioning is considered a luxury. Heating, however, is a necessity. Heat pumps were invented by people who understood air conditioning really well, but who did not have very much to do with their time. Their thinking went something like this: if we can take heat from a house and throw it outside when we want to keep the house cool, why not run it backwards in the winter taking heat from outside and throwing it inside?

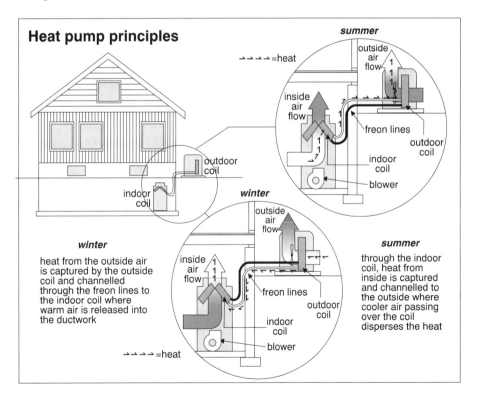

Refrigerators Heat Kitchens They got some clues from equipment that was already working in the house. For example, the air coming off a refrigerator feels pretty warm, and so does the air coming off the outside of a window air conditioner. When it is 85°F outside the house and you want to keep it at 75°F inside the house, a window air conditioner grabs heat from the 75°F air and throws it out to the 85°F air.

SECTION TWO: HEAT PUMPS

Let's Put The Window Air Conditioner In Backwards

Let's say it was 50°F outside and we wanted to keep the house at 70°F inside. Could we turn the window air conditioner around (install it backwards) so that it would take heat from the outdoor 50°F air and dump it into the 70°F air inside the house? Theoretically, we could, although the equipment would not work very well because it is not designed to do this.

Collect Heat From Outside

The principle of a heat pump is the same. In much the same way that the refrigerator can grab heat from its cool interior and throw the heat out into the kitchen, a heat pump grabs heat from the outside air and dumps it into the house.

There Is Some Heat In Cold Air

How can we get heat from cold air? Even though the temperature is low, there is heat in the air. If we can put something even colder outside, the heat will flow out of the air into that colder **thing.**

That colder **thing** turns out to be the same refrigerant we use in the summer for the air conditioning.

Use The Air Conditioning System In The Winter

The people who designed, built and sold air conditioning equipment knew that it was fairly expensive. In many parts of North America, the equipment sits idle much of the year. If they could use the same equipment that cools the house in the summer, to heat it in the winter, wouldn't that be great! They found with some modifications, they could do just that.

3.2 THE MECHANICS

Don't Worry About Humidity

When we talked about air conditioning, we talked about two goals. The first was lowering the air temperature, and the second one was lowering the humidity. When we are talking about heating, we do not have to worry about humidity adjustment as a goal, because as we heat the indoor air, its ability to hold moisture increases. Humidity adjustment does create some problems for us elsewhere, but we will talk about those later. The real goal here is to grab heat from outdoors and bring it into the house.

We are all experts at grabbing heat from the house air and throwing it outside now, so let's see if we can change our thinking slightly and work the other way.

Cold, Low Pressure, Liquid Goes Into Outdoor Coil

If we want to grab heat from outdoors when it is 45°F outside, we have to pass a cold liquid through an **outside coil**. This sounds similar to what we did indoors during the air conditioning season. We are going to use an **expansion device** to create a low pressure, low temperature liquid entering the **evaporator coil**. This cold liquid will see the 45°F outside air being blown across the coil, and will pick up some heat from it.

Liquid Picks Up Heat And Boils

Again, the **latent heat of vaporization** is important as the liquid is boiled off to a low pressure, low temperature gas. The gas moves out of the evaporator coil and moves into the house. Now we have a cool, low-pressure gas coming into the house, but we want to get the heat out of the gas.

SECTION TWO: HEAT PUMPS

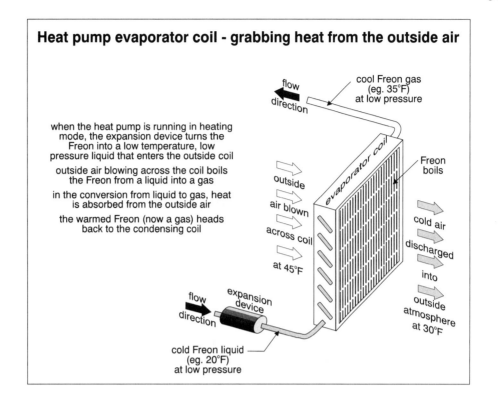

Heat pump evaporator coil - grabbing heat from the outside air

- when the heat pump is running in heating mode, the expansion device turns the Freon into a low temperature, low pressure liquid that enters the outside coil
- outside air blowing across the coil boils the Freon from a liquid into a gas
- in the conversion from liquid to gas, heat is absorbed from the outside air
- the warmed Freon (now a gas) heads back to the condensing coil

Cool Gas Compressed To Become Hot Gas

We go through a **compressor** (which can be indoors or out) to raise the pressure and temperature of the gas. We now have a hot, high-pressure gas that we pass through the **indoor coil**. In this situation, the indoor coil is the **condenser coil**. The hot gas passes through the coil and the house air blows across the coil. As the gas inside the coil is cooled, it condenses back to a liquid, giving off its heat to the house air.

Hot, High Pressure Becomes Cool, Low Pressure Liquid

This is how we grab heat from the outdoors and dump it indoors. To complete the Freon loop, we take the warm high pressure liquid back outside, passing it through an **expansion device** to create a cool, low pressure liquid that boils off in the outdoor coil (acting as an evaporator coil). Then we go through the cycle again.

SECTION TWO: HEAT PUMPS

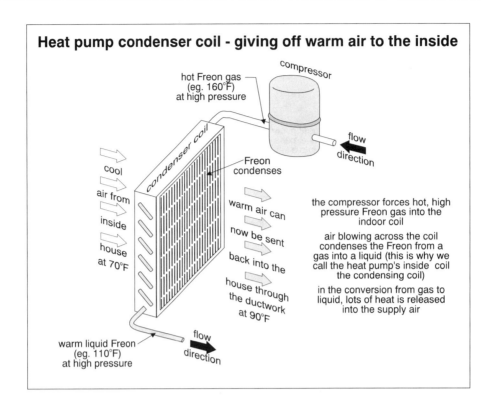

It's Reverse Air Conditioning

The process is almost the same as the air conditioning process, using a change of state and pressurizing and depressurizing the gas at the appropriate points to move heat from the outdoors to the indoors, in this case. With air conditioning, we move heat from the indoors to the outdoors. The condenser coil for air conditioning becomes the **evaporator** coil for heating. The **evaporator** coil for air conditioning becomes the **condenser** coil for heating.

Coils Reverse Their Roles

Reversing Valve

A **reversing valve** is used to change the direction of the Freon flow when changing from cooling to heating mode. The indoor and outdoor coils remain the same, the compressor remains the same and the expansion devices perform the same function, although there may be two instead of one. Depending on the climate and installation, both of the Freon lines may be insulated.

The electric reversing valve is controlled by the thermostat in the house. The reversing valve is typically located close to the compressor in a cabinet which, as we mentioned earlier, might be located indoors or outdoors.

Refrigerant Lines The Same

The liquid refrigerant line is still always the liquid line, and the suction line (gas line) is always the suction line.

SECTION TWO: HEAT PUMPS

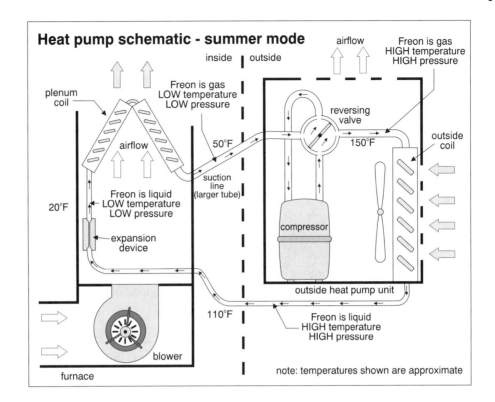

Large Temperature Difference Between Indoors And Out

When we talked about central air conditioning, we talked about trying to keep the house temperature 15°F cooler than the outdoor temperature. With a heat pump, we would like to keep the house considerably warmer than the outdoor temperature, particularly as the temperature falls. In the heating season, we are usually shooting for an indoor temperature of around 70°F. The outdoor temperature, depending on climate, may drop down to 0°F or below. This is a 70° differential between indoors and outdoors.

The ability of an air conditioner or a heat pump to move heat is reduced as the temperature outdoors gets further away from the desired indoor temperature. On a very hot summer day when the temperature is up around 100°F, most residential air conditioners cannot keep the house at 75°F. While we could design air conditioners to keep the house cool under these circumstances, they would be oversized for most conditions.

Heat Pumps Have To Work Harder

Heat pumps have a much tougher challenge, because while it is unusual to have a summer outdoor temperature that is 30°F higher than 75°F, it is not at all unusual to have an outdoor temperature in the winter that is 30°F to 50°F cooler than what we would like inside the house.

Heat Pumps Rated At 47°F and At 17°F

It is easy to understand how the colder it is outside, the harder it is to grab heat from the outdoor air. The amount of heat that a heat pump can deliver to a house varies with the outdoor temperature. It is common to rate heat pumps at two stages by the amount of heat they can deliver to the house when the outdoor temperature is 47°F and when it's 17°F. This gives two different capacity ratings. This is simply a convention that allows people to size the heat pump and to compare various systems.

SECTION TWO: HEAT PUMPS

Sizing Heat Pumps Based On Cooling Load
We mentioned earlier that heat pumps use the same equipment (same coils, same compressor, same Freon lines, etc.) for air conditioning and heating. The challenge, particularly in the northern climates of North America, is to size the equipment properly. We usually size the system based on **cooling loads** rather than **heating loads**. This means in northern climates, the heat pump cannot deliver enough heat during the cold winter weather, because of the big temperature differentials between inside and outside.

Swamp Like Environments
Why don't we make the heat pump large enough to heat the house on the coldest day? Because the air conditioning would not work well. If the air conditioner is dramatically oversized, it cools the house very quickly and does not pass air over the evaporator coil enough times to dehumidify it. You end up with a cold damp house, which is not a very pleasant environment. Further, the short-cycling of the air conditioner is hard on the compressor and compressor motor. Poor dehumidifying performance and shortened life expectancy are the primary reasons that many heat pumps are sized for the cooling load of the house, rather than the heating load, in colder climates.

Two-Stage Compressor
Modern two-stage compressors that operate differently in cooling and heating modes help to overcome this problem. The heat pump can operate at higher capacity in the winter than in the summer.

Oversizing
Air conditioning can be **slightly** oversized without too much trouble. Some experts say that oversizing by 25% to 35% of the cooling load can be done safely.

Temperature Rise Is Low
A heat pump operating in the heating mode may have a house air temperature rise of 20° to 30°F across the indoor coil. Temperatures at registers may be only 85° to 90°F.

Comfort Issue
Some people feel this is a cool draft, and don't feel comfortable with heat pumps as a result.

450 Cfm Per Ton
The recommended house airflow rate across a heat pump coil is roughly 450 cubic feet per minute for each ton. This is much more than a conventional furnace needs.

3.3 CO-EFFICIENT OF PERFORMANCE (COP)

Heat pumps typically use electric energy to drive the compressor and move air across the coils. While a heat pump is grabbing some heat from the outdoor air and delivering it into the house, it is consuming electrical energy. At some outdoor temperature, the amount of heat delivered to the house is just equal to the energy used to capture that heat. At this point, the **coefficient of performance (COP)** of a heat pump is 1.0.

COP Is 1.0 When Cost Equals Benefit
Let's put it another way. The COP is greater than 1.0 when we get more energy from outdoors than it costs us to collect the energy. The COP is less than 1.0 when we get less energy from outdoors than we spend to get it. The COP is 1.0 when the cost equals the benefit.

SECTION TWO: HEAT PUMPS

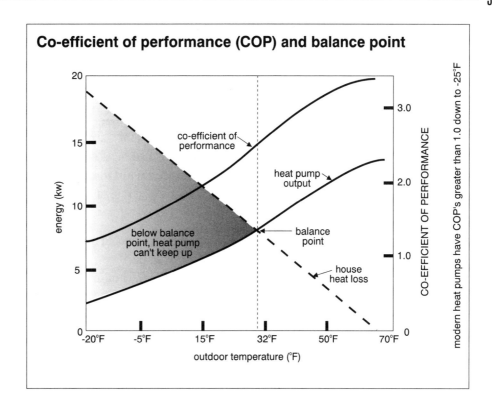

Co-efficient of performance (COP) and balance point

COP = 3.0 At 47°F

Many heat pumps operate at a coefficient of performance (COP) of approximately 3.0 when the outdoor air is 47°F. This means the heat pump is delivering three times as much energy as it is using. While many heat pumps can operate and capture some heat from the outdoor air when the outdoor temperature is below 0°F, it does not make sense to keep the heat pump running when the COP is less than 1.0. Some units have a COP that doesn't drop down to 1.0 until the temperature drops to -13°F! A COP of 1.5 is common at 0°F.

When COP Is Low Shut Off Heat Pump

When the COP is less than 1.0, it is costing more to bring heat into the house than it would cost to shut down the heat pump and provide electric heat directly into the house.

A heat pump operating at a COP below 1.0 consumes more than one unit of energy for every unit delivered. Since electric resistance heat operates at a COP of roughly 1.0 (it's roughly 100 percent efficient), it is better to shut off the heat pump and use electric heat when the COP drops below 1.0.

If Back-up Heat Fuel Is Cheaper

If the back-up heat source is a gas furnace, it may make sense to shut off the heat pump sooner. Let's say the gas furnace delivers heat at 50 percent of the cost of electricity, based on the fuel costs and efficiencies of the systems. Since electricity costs twice as much as the gas heat, any time the COP is below 2.0, the gas system is less expensive to operate, and the heat pump should be shut down!

However, heat pumps often shut down with COPs above 1.0 because even though they can capture some heat cost effectively, they can't provide enough heat to keep the house warm.

3.4 WHEN THE HEAT PUMP CANNOT KEEP UP

Heat Pump Stays Ahead

Sometimes houses need extra heat to keep them warm enough during the heating season. This is determined by a number of things, but boils down to this. If the heat pump can add heat faster than it is being lost through the walls, ceilings, etc., the heat pump can keep the house warm. The rate of heat loss from the house is a function of such things as –

1. the temperature difference between indoors and outdoors
2. the insulating value of the house envelope
3. the amount of air leakage into or out of the house

If the house loses heat faster than the heat pump can add it, we need some help.

Balance Point

The **balance point** of a heat pump is the temperature at which the amount of heat that the heat pump can collect from the outdoor air is exactly equal to the amount of heat being lost to the outdoors through the house. At the balance point, the heat pump would run constantly and the house temperature would stay at 70°F.

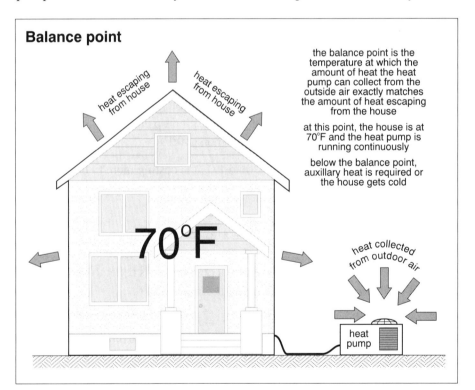

Heat Pump Falls Behind

As the outdoor temperature falls below the balance point, the heat pump can still add heat to the house, but heat is being lost faster than it is being added. The house temperature will fall. The heat pump needs help to keep the house comfortable, and auxiliary heat is introduced.

SECTION TWO: HEAT PUMPS

Back-up Heat

The auxiliary (back-up) heat for a heat pump can be electric resistance elements, gas furnaces, or oil furnaces, for example. When the auxiliary heat is a fossil fuel, the heat pump has to shut off when the furnace comes on. This is known as an **add-on system.**

With electric heat, the heat pump can keep working, and the electric elements can add the necessary amount of heat. This is known as an **all-electric system.**

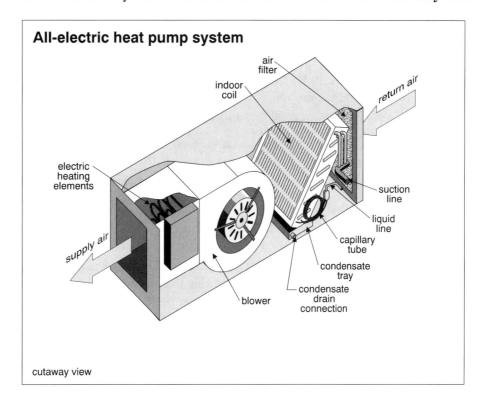

It is nice to be able to have the heat pump working and the auxiliary heat working at the same time. With electric heat as the back-up, whenever the COP is greater than 1.0, the heat provided by the heat pump is a good value. If you are getting heat from the heat pump at good value, but cannot keep up with the rate of heat loss from the house, the back-up electric heat can kick in and make up the shortfall.

Heat Pump Plus Electric Heat

When the outdoor temperature falls, and the COP drops below 1.0, it makes sense to shut the heat pump off and rely exclusively on the electric heat. This assumes the capacity of the electric system is adequate to heat the entire home.

Fossil Fuel Back-Up

With a fossil fuel system (gas or oil), we have to shut the heat pump off when the furnace comes on, because of the temperature differential across the heat pump coil. We'd like to have the furnace just help out, but we can't. Let's look at why.

The indoor coil (which in the winter is acting as a condenser coil) is located downstream of (or after) the furnace heat exchanger. Heat pumps are designed for air coming into the indoor coil at room temperature (70°F). When a conventional gas or oil furnace is working, the air coming off the furnace heat exchanger is 130°F or 140°F.

Furnace Air Too Hot For Coil

Under these circumstances, the air is so hot that it can't pick up any heat from the coil, and the heat pump doesn't work. When the heat pump sees 70°F air coming into the indoor coil, the air usually leaves at 85°F to 100°F. The 140°F air coming into the indoor coil is just too hot for the heat pump to add any more heat.

Can We Move The Coil?

The obvious answer is to put the indoor coil **upstream of (before)** the furnace heat exchanger, before the house air gets heated up from 70°F to 140°F. This would work great in the heating season, because then the heat pump would see 70°F house air and heat it up a little bit. The furnace could then heat the air the rest of the way when it was very cold outside. This sounds like a solution!

We'd Rust Out Furnace

In practice, this is not done. Would you put the indoor coil upstream of (or before) the heat exchanger, you would rust out the furnace heat exchanger when the system works as an air conditioner in the summer.

Why?

Remember that the indoor coil (evaporator coil during the summer) dehumidifies the house air as it passes across the coil. The 55°F or 60°F air that comes off the indoor coil during the summer is dry, but another problem is created.

Condensation Forms On Combustion Side Of Heat Exchanger

This cool, dry air would pass over the **house side** of the furnace heat exchanger, cooling the air on the **combustion side** of the heat exchanger. This air on the combustion side of the heat exchanger has not been dehumidified, and as we cool this air, we would create condensation. This condensation would accumulate on and rust the heat exchanger during the summer months when the furnace is not working.

This would dramatically shorten the life of a furnace, and may create an unsafe situation as the heat exchanger rusts through. As a result, the heat pump coils are always placed **downstream of (after) the heat exchanger**, after the air has passed over the heat exchanger.

Not An Issue With Electric Back-Up

This problem is only true of gas, oil or propane back-up heat. If the back-up heat is electric, there is no heat exchanger to rust. The electric heat elements can go after (downstream of) the heat pump coil. Both the heat pump and electric back-up can work at the same time.

3.5 HUMIDITY IS THE ENEMY

Frost On Outdoor Coil

We mentioned earlier that humidity can be an enemy of heat pumps. In the winter, the outdoor air temperature is low and the relative humidity is typically high. When the heat pump grabs heat from the outdoor air, the air moving across the outdoor coil gets cooler as it passes over the coil. Outdoor air at 35°F and 80% relative humidity goes into the coil. The air coming off the coil is at 20°F and 100% humidity. This air is saturated. At these temperatures, condensation doesn't form on the coil, but frost does. The frost that builds up on the coil inhibits air movement and heat transfer across the coil, making it difficult for the heat pump to do its job.

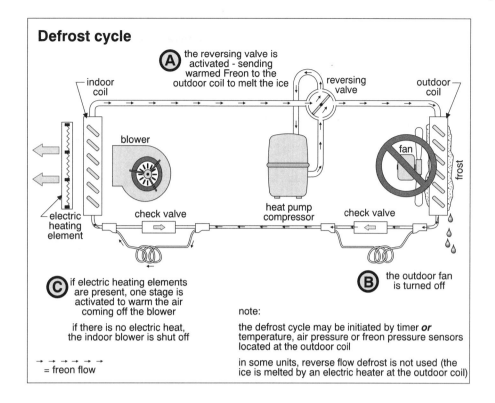

Defrost Cycle

As a result, heat pumps have **defrosters**. Some defrosters are electric heating elements located at the outdoor coil. Other defrosting systems turn the reversing valve around so that the heat pump temporarily acts as an air conditioner, bringing some heat from the house outside to melt the frost on the coil.

Timed Or Demand Defrost

The defrost cycles on some units are **timed**, defrosting every 30 to 90 minutes for example. On more sophisticated and energy efficient systems, the defrost is on a **demand** basis, where the unit senses a frost build-up and defrosts as necessary.

3.6 COST EFFECTIVENESS

In many northern climates, the cost effectiveness of heat pumps has been questioned when all factors in their life cycle costing are considered. Many heat pumps have been abandoned for heating, and are now simply used as air conditioning systems.

3.7 ONE LITTLE TWIST

Water-Source Heat Pumps

Just as we talked about air conditioners dumping house heat into water or the ground instead of the air in some cases, heat pumps can do the same thing. The deep water from wells, lakes, etc., is a constant 40°F to 50°F year round. This means that in the winter, we can capture heat from this water and bring it into the house. In the summer we can dump heat from the house in the water using **water-source heat pumps**.

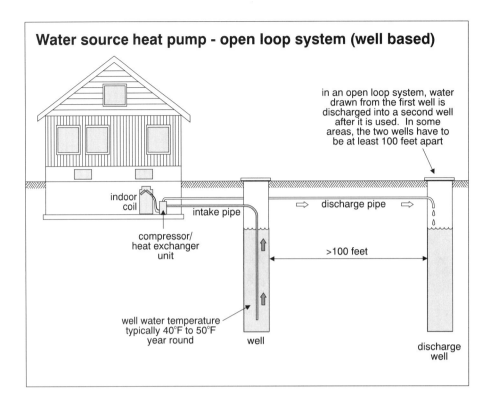

Ground-Source Heat Pumps The ground temperature is constant a few feet below the surface, year round. Again, we can steal heat from the ground in the winter, and dump heat into the ground in the summer. Underground piping systems are used in these **ground-source heat pumps**.

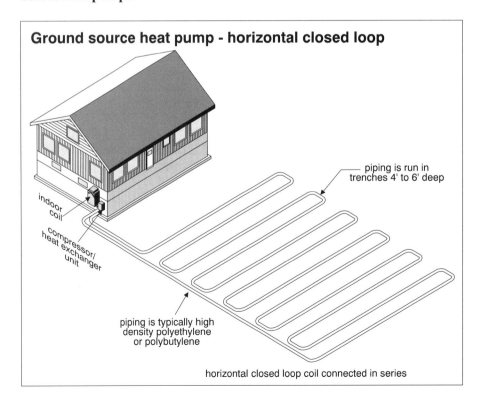

SECTION TWO: HEAT PUMPS

Air Conditioning & Heat Pumps
MODULE

QUICK QUIZ 1

☑ INSTRUCTIONS

- You should finish Study Session 1 before doing this Quiz.
- Write your answers in the spaces provided.
- Check your answers against ours at the end of this Section.
- If you have trouble with the Quiz, reread the Study Session and try the Quiz again.
- If you did well, it's time for Study Session 2.

1. The indoor coil on heat pump acts as an evaporator coil in the summer and a _____ coil in the winter.

2. The outdoor coil acts as a condenser coil in the summer and a _____ coil in the winter in a conventional air-to-air heat pump system.

3. The compressor on a heat pump can be –
   a. indoors
   b. outdoors
   c. all of the above

4. Where is the reversing valve is typically located?
   _____

SECTION TWO: HEAT PUMPS

5. Heat pumps may see a temperature difference between outdoors and indoors that is –

   a. smaller than what an air conditioner would typically see
   b. the same as what an air conditioner would typically see
   c. greater than what an air conditioner would typically see

6. Heat pumps are generally sized based on the –

   a. heating load
   b. cooling load
   c. other

7. The implication of an oversized heat pump during the summer months is –

   _____
   _____

8. Define the coefficient of performance.

   _____
   _____
   _____

9. What is a typical heat pump coefficient of performance when the outdoor temperature is 47°F?

   _____

10. At what COP level should the electric back-up heat be turned on and the heat pump turned off?

    _____

11. Define balance point with respect to heat pumps.

    _____
    _____

12. An auxiliary electric heating system is typically installed upstream of (before) the heat pump coil.
    True ☐ False ☐

13. A gas furnace back-up system typically has the heat exchanger upstream of (before) the heat pump coil.
    True ☐ False ☐

SECTION TWO: HEAT PUMPS

14. Explain why the location is important for each back-up system.

    a. Electric heat

    _____

    _____

    b. Gas Furnace

    _____

    _____

15. Why does frost sometimes develop on the outdoor coil of a heat pump?

    _____

    _____

16. Name two types of heat pumps other than air-to-air.

    _____

    _____

If you didn't have any difficulty here, then you are ready for Study Session 2.

Key Words:
- *Air conditioning in reverse*
- *Outside coil*
- *Indoor coil*
- *Expansion device*
- *Compressor*
- *Refrigerant*
- *Latent heat of vaporization*
- *Reversing valve*
- *Condenser coil*
- *Evaporator coil*
- *Coefficient of performance*
- *Balance point*
- *Back-up heat*
- *Frost*
- *Water-source heat pump*
- *Ground-source heat pump*

SECTION TWO: HEAT PUMPS

Air Conditioning & Heat Pumps
MODULE

STUDY SESSION 2

1. You should have finished Study Session 1 and Quick Quiz 1 before starting this Study Session.

2. This Study Session deals with heat pump components and heat pump problems.

3. At the end of this Study Session, you should be able to –

    - List the components of heat pumps
    - List the differences between heat pumps and air conditioners
    - Explain the defrost cycle in three sentences
    - List nine tricks for identifying heat pumps
    - List seven common heat pump problems

4. This Study Session should take you roughly one hour to complete.

5. Quick Quiz 2 is included at the end of this Session. Answers may be written in your book.

SECTION TWO: HEAT PUMPS

Key Words:

- *Compressor*
- *Evaporator coil*
- *Expansion device*
- *Condenser coil*
- *Outdoor fan*
- *Indoor fan*
- *Condensate tray*
- *Condensate lines*
- *Refrigerant lines*
- *Evaporator fan*
- *Duct system*
- *Air filter*
- *Thermostat*
- *Heat sink*
- *Heat source*
- *Reversing valve*
- *Emergency heat*
- *Accumulator*
- *Timed defrost*
- *Demand defrost*
- *Oversized*
- *Undersized*
- *Inoperative*
- *Poor location*
- *Iced-up coil*
- *Weak airflow*
- *No back-up heat*

SECTION TWO: HEAT PUMPS

▶ 4.0 HEAT PUMPS IN PRACTICE

4.1 UNDERSIZED OR OVERSIZED

Design For The Cooling Load

As we've discussed, the size of the heat pump that would be ideal for cooling is not always the same as the ideal heating size. If you are in a climate where the maximum heat gain on a hot summer day is similar to the maximum heat loss on a cold winter day, then sizing a heat pump is easy. You simply design it for the same heating and cooling capacity. If the heating load is lower than the cooling load, you design it for the cooling load.

Cooling Load

Even if the heating load is larger, most heat pumps are still designed for the cooling load because if the unit is too large for cooling, it will make the house cool and damp, which is very uncomfortable. We talked about this earlier.

Two Stage Compressors

One exception is two-stage compressors which allow the heat pump to operate at different capacities. This can allow a heat pump to operate at lower capacity during the cooling season and higher capacity during the heating season.

We've talked about typical heating and cooling loads elsewhere, and you should use those same rules of thumb here. As we've discussed, the typical cooling load in the southern United States is one ton for every 450 to 700 square feet of living space, whereas in moderate and cooler climates, one ton is adequate for 700 to 1,000 square feet.

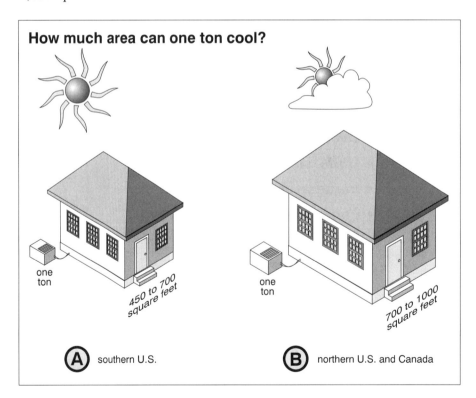

How much area can one ton cool?

Ⓐ southern U.S. — one ton, 450 to 700 square feet

Ⓑ northern U.S. and Canada — one ton, 700 to 1000 square feet

SECTION TWO: HEAT PUMPS

1 To 5 Tons   Common residential heat pump sizes range from one to five tons of cooling. Generally speaking, less cooling is required in dryer climates, and two or three story houses need less cooling per square foot of living space than bungalows.

As a rough rule for heating, 40 to 60 BTUs per square foot is recommended for poorly insulated, leaky homes in northern climates. Twenty-five to 40 BTUs per square foot is recommended for better insulated, tight homes.

4.2 HEAT PUMPS ARE SIMILAR TO AIR CONDITIONERS

Heat pumps have many of the same components as air conditioners. This makes sense because they are air conditioners during the cooling season. These components can experience the same problems as air conditioners.

1. Compressor
2. Indoor coil
3. Metering (expansion) device
4. Outdoor coil
5. Outdoor fan (condenser fan)
6. Condensate tray and condensate lines
7. Refrigerant lines
8. Indoor fan (evaporator fan)
9. Duct system
10. Duct insulation
11. Air filter
12. Thermostat

Water-To-Air Heat Pumps   Just as we can have water-cooled air conditioners, we can also have water-cooled (and heated) heat pumps. In this case, the water is used as a **heat sink** in the summer (a place to dump heat we've collected from the house). It can also be used as a **heat source** in the winter (collecting heat from the water and dumping it into the house). Ground-source heat pumps are also found. We touched on these earlier.

4.3 DIFFERENCES BETWEEN AIR CONDITIONERS AND HEAT PUMPS

The differences include the following components and functions:

Reversing Valves
1. A heat pump has a **reversing valve**. This allows us to change the direction of the refrigerant flow through the system.

Coils Change Their Functions
2. The indoor and outdoor coils reverse function summer and winter. On heat pumps (and air conditioners) the indoor coil is the evaporator coil in the summer. The outdoor coil is the condenser coil in the summer. With heat pumps in the winter, the **indoor coil** acts as the **condenser coil** and the **outdoor coil** acts as the **evaporator coil**.

SECTION TWO: HEAT PUMPS

Both Lines Insulated

3. On heat pump systems, **both** of the **Freon lines** may be **insulated**. On air conditioning systems, only the suction line (gas line) is insulated.

Outdoor Air Temperature Sensor

4. Heat pumps may include an **outdoor air temperature sensor (ambient stat)**. This may tell the heat pump to shut off when it gets too cold outside. It may also be arranged to lock out the back-up heat above a certain outdoor temperature.

Emergency Heat Setting

5. Heat pumps often have an **Emergency Heat** setting on the thermostat. This allows you to switch over to back-up heat if the heat pump isn't working properly.

Two-Stage Thermostat

6. Many heat pumps have **two-stage thermostats**. The first stage brings on the heat pump, and if the temperature continues to fall, the second stage brings on the back-up heat. There are several variations on this approach.

Indoor Compressors

7. The compressor does not have to be outdoors on a heat pump. The **compressors** are often indoors. These systems are called **triple-split** systems. The compressor and reversing valve are in cabinet, usually near the indoor fan coil unit.

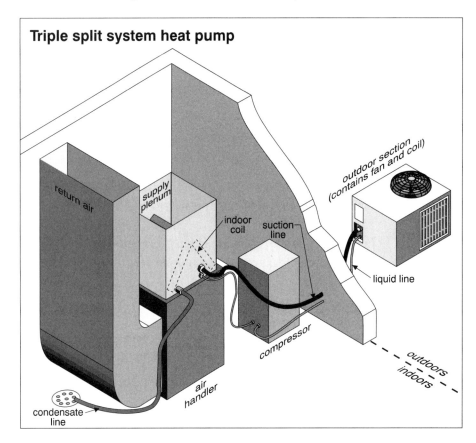

There are some advantages to putting the compressor indoors.

- We probably don't need a crankcase heater.
- Heat generated by the mechanical action of the compressor is kept indoors. (good in the heating season, not so good in the cooling season)

SECTION TWO: HEAT PUMPS

- It is easier and more pleasant to service an indoor unit in severe weather conditions.
- The equipment and their controls see a dryer, more stable environment and may enjoy a longer life.

Two Expansion Devices

8. Heat pumps often use **two expansion devices**. Air conditioners use only one. With heat pumps, there may be two expansion devices on the liquid line, one just upstream of each coil. Check valves and a bypass line are used to step around expansion devices that are not in use during that mode. In the cooling mode, the expansion device close to the indoor coil will be active. In the heating mode, the expansion device near the outdoor coil will be active.

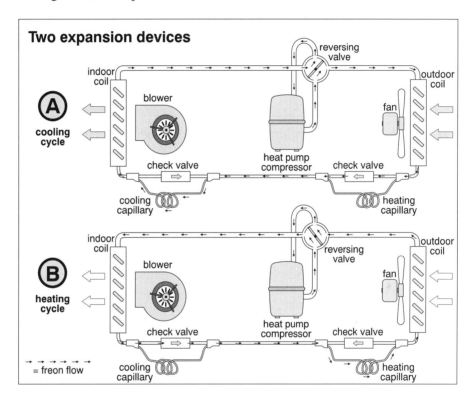

Defrost Cycles

9. Heat pumps have **defrost** mechanisms. These may be electric heaters, or they may use a reversal of cycle, controlled by temperature sensors, air pressure sensors, Freon pressure sensors or timers. Air conditioners do not need defrost mechanisms.

Accumulators

10. All heat pumps have **accumulators**. This is a device that collects liquid just upstream of the compressor. In the winter, the evaporator coil (outdoor coil) cannot always boil all of the liquid off into a gas. There is more refrigerant than we need in the system. We don't want liquid going into the compressor because it could damage it. The accumulator collects any liquid, allowing only the gas to flow through to the compressor. Some air conditioners have accumulators, but virtually all heat pumps do. The accumulator is usually located between the reversing valve and the compressor.

SECTION TWO: HEAT PUMPS

High Compression Ratios On Heat Pump Compressors

11. Compressors for heat pumps operate at lower outdoor temperatures, because they are operating in the winter. Their **compression ratios are much higher** than the three or **four to one** seen by air conditioners. Compression ratios for heat pump compressors can be greater than **eight to one**. This puts much more strain on the compressor. Heat pump compressors are typically better built than air conditioning compressors.

Crankcase Heaters

12. Crankcase heaters are often found on air conditioners, but are optional. **Crankcase heaters** are required on heat pumps. They can be external or internal to the compressor.

Freon Line Temperatures Different

13. In an air conditioning system, the suction Freon line feels cold and the liquid line feeds warm when the system is operating during the summer. When a heat pump is operating in the winter, the suction and liquid line temperatures can be confusing. We'll look at these shortly.

4.4 HEAT PUMPS DIFFER FROM GAS AND OIL FURNACES

Must Move Lots Of Air

The volume of air that has to move through a heat pump system is large compared to conventional furnaces. An airflow of approximately 450 cubic feet per minute for every ton is typically required. Heat pumps often suffer because of inadequate airflow.

Lower Discharge Temperature

With a gas or oil furnace, it's not unusual to have 130°F or 140°F air at the supply registers. With a heat pump, 85°F to 100°F air is more common. The relatively cool air causes many people to think the system isn't working, when in fact, this is appropriate. It is a mistake to reduce the rate of airflow through the system to raise the temperature. This puts more load on the compressor and may lead to compressor failure.

Do Not Operate In The Heating Mode When Over 65°F

Heat pumps should never be operated in the heating mode when the outdoor temperature is above 65°F. This may damage the compressor.

4.5 DEFROST CYCLE

Let's review our discussion on defrosting and take it a little farther.

Why Is Defrosting Necessary?

During the winter, the outdoor air is relatively cool with high relative humidity, typically 70% or 80%. Because the air is so cold, it can't hold much water vapor. Consequently, it doesn't take much moisture to almost saturate the air. The colder air is, the less moisture it can hold. Dropping the air temperature even slightly will cause the moisture to fall out of the air as water or frost.

SECTION TWO: HEAT PUMPS

When we use a heat pump in the heating mode, the outdoor coil **lowers** the temperature of the outdoor air passing across it. This often causes the relative humidity to go up from 80% to 100%. Condensation or frost forms on the coil. If more than roughly one-eighth inch of frost is allowed to collect, this will obstruct the air flow and inhibit heat transfer, effectively making the unit useless. The defrost cycle is needed to melt this ice before it can build up to the point where it inhibits operation.

Frost accumulates on the outdoor coil when the coil temperature approaches 32°F.

How Defrosting Is Done

There are two ways that heat pumps get rid of the frost that accumulates in the winter. Some use electric heaters, while others activate the reversing valve to go into the cooling mode. In either case, when the system is in the defrost cycle, the outdoor fan shuts down.

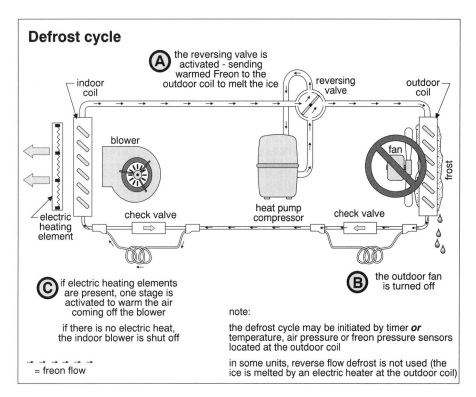

In some systems, the indoor back-up electric heat will kick in so that the occupants of the house don't feel cool air. The indoor house fan keeps blowing air across the indoor coil while the outdoor fan is shut down.

With an add-on heat pump (which has a gas or oil furnace), the indoor fan may also shut off during the defrost cycle, to avoid blowing cool air around the house.

Controlling The Defrost

Many systems have a **time-and-temperature defrost** mode. In effect, the heat pump checks the outdoor temperature near the coil every 30 to 90 minutes (depending on how it is set up).

Time-and-temperature Defrost

If the temperature is at or below freezing (32°F), the system will go into defrost mode for up to ten minutes. This is a somewhat rudimentary defrost cycle.

Demand Defrost

Many consider **demand defrost** to be superior. This may be triggered by measuring the air pressure drop across the coil, the coil temperature, or the Freon pressures moving through the coil. The defrost cycle may be as short as one minute. This is a more sophisticated system, but is also more complex and expensive.

Concerns With Defrosting

- The need to defrost reduces the efficiency of heat pumps, particularly in climates that are cold enough to need frequent defrosting.
- It's not always easy to get rid of the melted frost without forming ice dams that obstruct the airflow around the coil.
- Reversing the direction of refrigerant flow to defrost the coil can be hard on the compressor.

Heat Pumps Off The Ground

The water melted from the coil must drain freely away, or it will refreeze. Heat pumps should be well off the ground. This also helps keep drifting or deep snow from obstructing the heat pump.

4.6 IDENTIFYING HEAT PUMPS

At first glance, it can be difficult to differentiate heat pumps from air conditioners, especially during the summer months. The following are some of the ways to tell.

1. Look at the data plate. It may say **Heat Pump** on it, or the model number may start with **HP**. If it does not, you can jot down the manufacturer's name and model number and contact the manufacturer.
2. If the thermostat has an **Emergency Heat** setting, you can be sure it's a heat pump.
3. If you take the cover off the thermostat (beyond the Standards) and find that it's a two- stage thermostat, this indicates a heat pump. (You must know what you're looking at here.)
4. If both Freon lines are insulated, it's a heat pump.
5. If you open the condenser cabinet and find a reversing valve, it's a heat pump.
6. If you find two expansion devices with bypasses, it's a heat pump.
7. If the compressor's indoors and it's an air to air system, it's a heat pump.
8. If there is an outdoor thermostat connected to the control wiring, it's probably a heat pump.
9. If it's winter and the unit is operating, it's a heat pump.

Is It Still Being Used As A Heat Pump?

In many northern climates, heat pumps were often installed with the encouragement of local utilities. For various reasons, they were found to be unsatisfactory. Many have been disconnected as heat pumps and are used only as air conditioners. You won't be able to determine this visually, but your client may want to ask the current owner whether the heat pump is still used for heating. In many cases, it is cheaper to have the back-up heat supply all of the heating, especially if the back-up is a high-efficiency gas furnace. Where the available back-up heat is electricity, heat pumps are more cost effective and more likely to be in service, even in colder climates.

SECTION TWO: HEAT PUMPS

If you're looking at the house in the winter, and the system is operating on the **Emergency Heat** setting, chances are the heat pump is not operative.

Freon Temperatures In Heating Mode

A heat pump operating when the outdoor temperature is 35°F will have normal Freon (R-22) line temperatures in the following ranges:

• Liquid line upstream of expansion device - 105 to 115°F
• Liquid line downstream of expansion device - 15 to 20°F
• Suction line upstream of compressor - 25 to 30°F
• Suction line downstream of compressor - 160 to 170°F

Efficiencies

Seasonal Energy Efficiency Ratio

The **Seasonal Energy Efficiency Ratio** (SEER) of an air conditioner is simply a ratio of how many BTUs per hour you're getting out of the system relative to the watts of electrical energy consumed to run the unit.

$$\text{SEER} = \frac{\text{Total Cooling Output Over Season}}{\text{Total Electrical Input Over Season}}$$

SEER ratings of 6 are typical for old air conditioners. New split-system air conditioners (since 1992) are typically at least 10, and high-efficiency air conditioners are typically about 14. While some clients may ask, most inspectors do not report on efficiencies, and the Standards don't require it.

Heating Seasonal Performance Factor

Heat pumps are rated using SEER for their cooling efficiencies. Their heating efficiency is measured using HSPF (**Heating Seasonal Performance Factor**).

$$\text{HSPF} = \frac{\text{Total Heating Output Over Season}}{\text{Total Electrical Input Over Season}}$$

Ratings of 6.5 are average, with high efficiency systems hitting 9.0.

4.7 CONDITIONS

All of the problems associated with air conditioning components may also be experienced with heat pumps. Let's review quickly the common problems with these components.

Compressor
- Excess noise/vibration
- Short cycling
- Running continuously
- Out of level
- Excess electric current draw
- Wrong fuse or breaker size
- Electric wires too small
- Missing electrical shut off
- Inoperative
- Inadequate cooling

Condenser
- Dirty
- Damaged
- Corrosion
- Clothes dryer or water heater exhaust too close

Evaporator
- Temperature split too high
- Temperature split too low
- No access to coil
- Dirty
- Frost
- Top of coil dry
- Corrosion
- Damage

Condensate System
- Pan leaking or overflowing
- Dirt in pan
- Pan cracked
- Inappropriate pan slope
- Rust or holes in pan
- Pan not well secured
- No auxiliary pan
- No float switch

Condensate Drain Line
- Leaking/damaged/split
- Disconnected/missing
- Blocked or crimped
- No trap
- Improper discharge point

SECTION TWO: HEAT PUMPS

Condensate Pump
- Inoperative
- Leaking
- Poor wiring

Refrigerant Lines
- Leaking
- Damaged
- Missing insulation
- Lines too warm or too cold
- Lines touching each other
- Low points or improper slope in lines

Expansion Device
- Capillary tube crimped/disconnected/leaking
- Thermostatic expansion valve loose/clogged/sticking

Condenser Fan
- Excess noise/vibration
- Inoperative
- Corrosion
- Mechanical damage
- Obstructed air flow
- Dirty

Evaporator Fan
- Undersized
- Misadjustment of belt or pulleys
- Excess noise/vibration
- Dirty
- Dirty or missing filter
- Inoperative
- Corrosion
- Damage

SECTION TWO: HEAT PUMPS

Duct System
- Undersized
- Incomplete
- Dirty
- Disconnected or leaking
- Obstructed or collapsed
- Poor support
- Poor balancing
- Humidifier damper missing
- Supply/return registers — too few
- Supply/return registers — poor location
- Supply/return registers — obstructed
- Insulation missing or incomplete
- Vapor barrier missing

Thermostats
- Inoperative
- Poor location
- Not level
- Loose
- Dirty
- Damaged
- Poor adjustment/calibration

There are some other conditions to watch for as well. The following additional problems are common on heat pumps:

1. Oversized for cooling and/or undersized for heating
2. Heat pump inoperative in heating or cooling mode
3. Poor outdoor coil location, including
    a. under the drip line of the roof
    b. where snow drifts accumulate
    c. where air re-circulates and is trapped in an enclosed area, or
    d. where the outdoor coil is exposed to the prevailing wind
4. Outdoor coil is iced up
5. Airflow problems
6. No back-up heat, or back-up heat does not work
7. Old

4.7.1 OVERSIZED FOR COOLING AND UNDERSIZED FOR HEATING

Almost all heat pumps in northern climates are undersized for heating. Their sizing should be determined by the cooling load. There is a temptation to oversize for cooling so that the heating load can be satisfied by the heat pump. As discussed earlier, this is usually poor practice.

Causes Many heat pumps have to be designed this way.

Implications Auxiliary heat is needed in the winter. In the summer, the cooling may produce a damp unpleasant indoor climate and the system may short cycle.

Strategy Since we're not going to do heat loss or heat gain calculations, your estimates of appropriate size will be very rough. Don't go out on a limb and condemn a unit as being undersized or oversized. You're wiser to describe it as suspect, and recommend further investigation.

Not A Perfect World Again, many people live with air conditioning systems that are compromises. Very few residential air conditioning systems are optimal design. Good home inspectors adjust client expectations to recognize this.

In northern climates, look for auxiliary heat to help the heat pump when it's cold.

4.7.2 INOPERATIVE IN HEATING OR COOLING MODE

Causes The system may not operate in the heating mode because –

- the system may have been disconnected in the heating mode intentionally
- there may be one of several mechanical or electrical problems
- the outdoor temperature may be lower than the heat pump lock-out temperature, or
- the thermostat may be in the Emergency Heat position.

If the system doesn't operate in the cooling mode, treat it as an air conditioner that doesn't work.

A service person should be contacted, since troubleshooting heat pumps is well beyond our scope.

Implications There may be no heat, or there may be complete dependence on the auxiliary heat. If the system doesn't work in the cooling mode, there will be no air conditioning and may be no heat.

Strategy Turn the thermostat up to see if the heat pump responds (assuming the outdoor temperature is below 65°F). Check that the heat pump is producing the heat, not the auxiliary heat.

Locked Out   If it is very cold outside, the heat pump may be locked out because the COP is too low to run the heat pump economically. It may also be locked out because we are below the balance point, and the heat pump can't keep up with the heat loss. Depending on the outdoor temperature, the type of back-up heat, and the set-up of the system, the heat pump may be operational but locked out.

If the system is set to the cooling mode and the outdoor temperature is above 65°F, the system should respond to lowering the thermostat. If it doesn't respond or doesn't deliver cool air, recommend further investigation. Let your client know that the system may not deliver any heat either, although you can't test that.

4.7.3 POOR OUTDOOR COIL LOCATION

Cause   This is an installation issue.

Implication   Inefficient operation, decreased comfort and increased costs will result if the outdoor coil is not well located.

Strategy   Watch for outdoor units –

- with restricted intake or discharge air flow
- within 6 feet of a clothes dryer, water heater vent or high efficiency furnace vent
- under the drip line of the roof
- below snow drift depth or with no provision for condensate to drain away freely

Some say the best location for a heat pump outside is on the east side of the house, but practical issues are probably more important.

We added furnace discharge vents to the list here. These vents are not an issue for air conditioning systems because the air conditioner and furnace would never operate at the same time. It can be an issue with heat pumps because heat pumps and furnaces may operate at the same time. The moisture in the furnace exhaust may ice up the heat pump coil quickly.

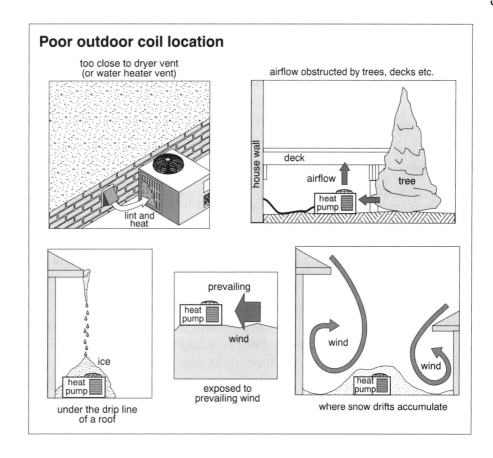

Poor outdoor coil location

Snow Accumulation Areas	Where snow tends to accumulate, the heat pump should be on a frame or stand several inches above the ground. This helps to prevent snow from blocking the air movement across the coil.

There should be 1 to 3 feet of clearance for the intake and 4 to 6 feet for the discharge side.

4.7.4 COIL ICED UP

Causes The outdoor coil may be iced up because –

- the defrost sensor is not working
- the control wiring has malfunctioned, or
- a heater has failed

Implication Poor heating or no heat is the implication.

Strategy Look for frost or ice on the outdoor coil.

SECTION TWO: HEAT PUMPS

4.7.5 AIRFLOW PROBLEMS IN HOUSE

Causes   This is typically a result of –

- dirty filters, fans or coils
- blocked supply or return registers
- fan speed too slow, or
- undersized or poorly balanced ducts

Implications   Poor heating and cooling is the implication.

Strategy   Perform your tests of the supply registers and return grilles as we have discussed previously.

Poor Register And Grille Location   Watch for supply registers and return grills that favor either heating or cooling. For example, ceiling supply registers are fine for cooling but are not very good for heating.

Low level returns are better for heating than for cooling. Ideal systems have high and low supply registers and return grilles. These are very rare. The best compromise for both heating and cooling is floor registers near outside walls below windows. Return grilles can be high and low with a damper on the low return grille. The damper is open in the heating season for low level return and closed in the cooling season allowing high level return.

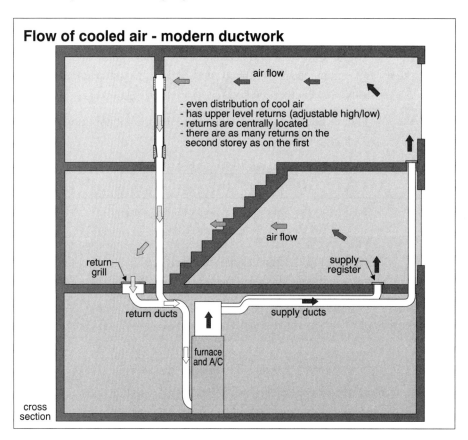

SECTION TWO: HEAT PUMPS

Ducts In Unconditioned Spaces
The more of the duct system that passes through unconditioned spaces (attics and crawlspaces), the worse the situation is. Comfort suffers and costs increase. Insulating and sealing ducts helps, but is never 100 percent effective. Warn your clients if there is much ductwork in unconditioned spaces.

4.7.6 BACK-UP HEAT PROBLEMS

The back-up heat may be inactive or ineffective.

Causes This may be because –

- elements are burned out
- there is a control wiring problem
- if it's a furnace, there may be mechanical problem, fuel problem or electrical problem, or
- if it's missing, it may be an installation issue or the heaters may have been removed. The heat pump system may be large enough to heat the entire home. Two-stage or double compressors may make the system variable capacity.

Implication The result is often inadequate heat in cold weather.

Strategy When looking at a heat pump, determine whether back-up heat is needed. You should know this from local installation practices. You can also approximate this by knowing the rough heat loss from the house and determining the heating capacity of the heat pump.

Back-Up Heat If back-up heat is needed, has it been provided? What type is it? If the back-up heat is electric, it can operate in conjunction with the heat pump. If the back-up heat is gas or oil, the heat pump has to be shut off when the gas or oil burner comes on.

Testing Back-Up Heat If the back-up heat is electric, it may be several staged elements that come on one at a time as required. You should, as part of your inspection, make sure that auxiliary heat is provided. You should also make sure that it works. Switching to the Emergency Heat setting, and using an ampmeter to measure the current draw through the auxiliary heating circuit is one way to accomplish this. If it is a staged system, you should make sure that each of the auxiliary electric elements comes on appropriately. Again, this goes beyond the Standards.

SECTION TWO: HEAT PUMPS

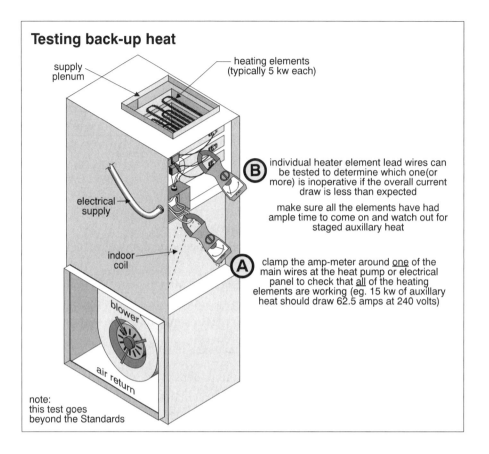

Temperature Limitations	You should avoid testing a heat pump in the heating mode above 65°F. Do not run a heat pump in the cooling mode below 65°F.
Back-Up Heat On Constantly	The electric back-up heat should only come on when the heat pump can't deliver enough heat. If the back-up heat is on with the heat pump when the temperature is mild (above 40°F), there may be a wiring or control setting problem. Recommend further investigation.
Five To Seven Minutes Between Cycles	Heat pumps, just like air conditioners, have a built-in time delay to prevent short on/off cycles which create a large pressure for the compressor to try to start up against. This can damage the compressor. Any time the heat pump is shut off, it probably won't recycle for five to seven minutes. A few manufacturers go even longer before allowing the system to re-start.

SECTION TWO: HEAT PUMPS

4.7.7 OLD

Heat pump compressors last 8 to 15 years, typically. Use the methods discussed in the Air Conditioning section to check the age of the heat pump and compressor.

We talked earlier about determining the age from the data plates. Remember, predicting life expectancies goes beyond the Standards.

Compressor Is Heart — The compressor is the single most expensive component of an air conditioning compressor. Replacement costs of $1,000 or more are not unusual. Compressors usually come with a five-year warranty, although some manufacturers offer a ten-year warranty.

4.8 OTHER TYPES OF HEAT PUMPS

We've been talking primarily about air-to-air (or air-source) heat pumps, where heat is captured from the outdoor air and delivered to the house air indoors. We've touched on some other systems. Let's review these and go a little further.

4.8.1 WATER-TO-AIR SYSTEM

In the heating season, heat is captured from water in wells, ponds, rivers or lakes and is transferred to the air in the house. During the summer, heat from the house is dumped into these bodies of water.

Pros And Cons — These systems are independent of outdoor temperatures and can have high seasonal efficiencies. There is no defrost cycle needed. These systems depend on good water quality and quantity. The biggest challenges are high installation cost, and disposal of the waste water in an environmentally acceptable way.

Open Or Closed Loop — **Water-source heat pumps** may be **open** or **closed** loop. Open-loop systems draw water from a source and discharge it somewhere else. For example, they may draw water from one well and dump it into another. Some authorities say these wells have to be 100 feet apart. Closed-loop systems circulate a liquid (typically an antifreeze such as propylene glycol, calcium chloride or methyl alcohol) through a loop submerged in a lake, for example. The piping is usually polyethylene or polybutylene, but not PVC. The joints are thermal fused. Mechanical fittings such as clamps have been unreliable.

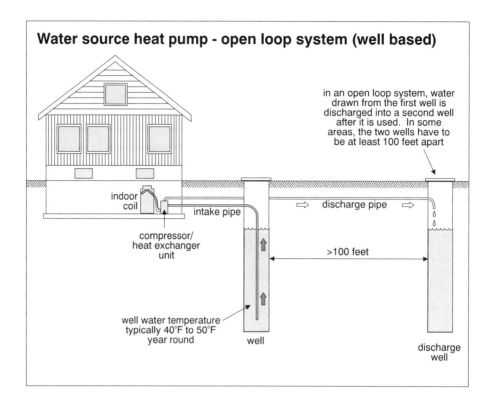

Lots Of Water These systems can use 5,000 to 10,000 gallons of water per day.

4.8.2 EARTH-TO-AIR SYSTEM

Ground-source heat pumps collect heat from the ground in the winter and deliver it into the house air. They have similar advantages to water-source heat pumps. In the summer, heat is collected from the house air and dumped into the ground.

The loops can be shallow, running horizontally 3 to 6 feet below the surface, or they may be vertical, going down as much as 200 feet below grade. The loops may be 300 to 600 feet long for each ton of capacity.

SECTION TWO: HEAT PUMPS

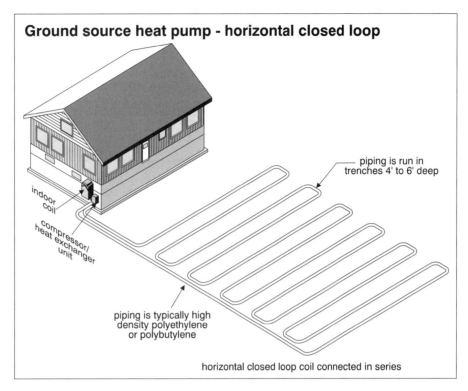

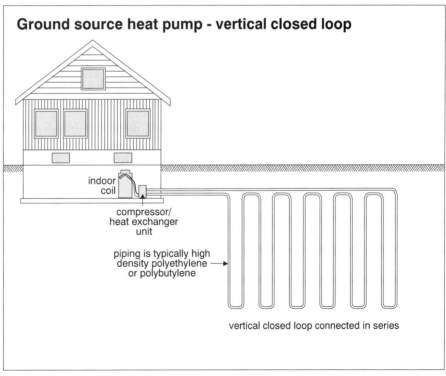

Components The indoor components for ground or water source heat pump typically include –

- a coil to transfer heat to or from the house air. This is similar to any heat pump or air conditioner
- refrigerant lines
- an expansion device
- a compressor
- a reversing valve
- a second coil, usually a water jacket that transfers heat between the refrigerant and the antifreeze or water
- a pump(s) to move the antifreeze or water

These systems may or may not provide all the heat needed for the home. Back-up heating is common. It is also common to use the heat pump to heat domestic water, helping the conventional domestic water heater.

In a vertical closed-loop system, the loops might be in series or parallel. The horizontal loop systems may need more pipe to keep the same heat transfer because at shallow depths, the soil temperature is not as uniform as it is deeper down.

Pipe Materials Pipe is usually either high-density polyethylene with heat-fused butt joints or polybutylene with heat-fused socket joints. PVC with solvent-welded fittings are no longer recommended for closed loops because the wide variation in temperature causes leaks. Copper piping is a good material, but it is usually too expensive. Also, copper piping may be attacked by some soil types.

Soil Shrinkage In some cases, ground-source heat pumps are hampered by soil that shrinks and pulls away from the pipe, severely reducing efficiency. In other cases, the soil acts as an insulator rather than a heat sink around the pipe, again reducing efficiency.

Expensive You should find out how effective and popular heat pumps are in your area and what types are used. Environmental issues restrict the use of water and ground source systems in some areas. These systems are expensive to install.

4.8.3 SOLAR SYSTEMS

Solar energy-to-air systems capture heat from the sun and transfer it to the house air. These specialized systems are beyond the scope of our discussion.

SECTION TWO: HEAT PUMPS

4.8.4 BIVALENT SYSTEMS

Heat Pump And Condensing Furnace

This type of system is a conventional heat pump with a condensing gas furnace built in. The system, made by a company called Kool-fire International, can use natural gas or propane. During the summer, it works just like an air conditioner. During the relatively mild part of the heating system, it functions like a conventional heat pump.

When the temperature drops below the balance point, something unusual happens. A gas burner mounted in the outdoor cabinet kicks on and the outdoor fan that blows air across the outdoor coil shuts down. In effect, the system stops working as a heat pump. The outdoor coil no longer tries to grab heat from the cool outdoor air. That is why the outdoor fan stops. The cabinet is not drawing cool outdoor air into the cabinet.

Burner Heats Outdoor Cabinet

The burner starts up and warms the interior of the cabinet to about 60°F. The outdoor coil transfers the heat from the air in the cabinet into the refrigerant. The refrigerant then carries the heat to the indoor coil where is it transferred to the house air.

Outdoor Coil Is Furnace Heat Exchanger

The outdoor coil becomes a heat exchanger much like the heat exchanger on a furnace. The outdoor cabinet is, in effect, a furnace cabinet. There is no chimney. Exhaust gases escape from a small opening in the top of the cabinet. The products of combustion are kept in the cabinet long enough to cool and condense. This results in very efficient heat transfer and when the system is operating in this mode, it is as efficient as a typical condensing furnace (90 percent or more efficient).

SECTION TWO: HEAT PUMPS

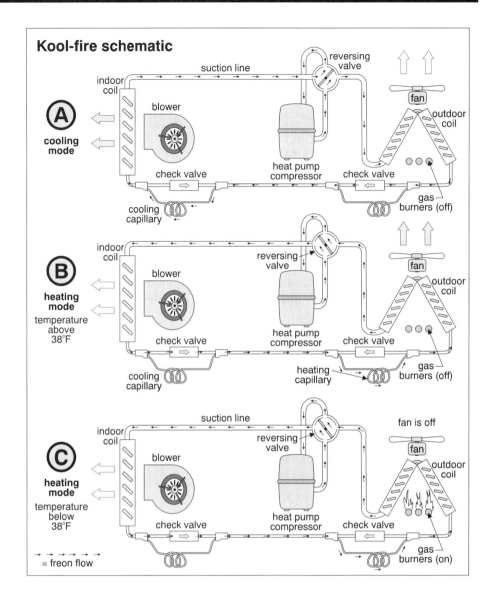

Condensation With all of the condensation in the cabinet, there is some concern about corrosion. The manufacturer claims that this is not a big problem, and some outside independent studies tend to support this. We have not seen enough field installations to be sure whether the high humidity levels and condensation within the cabinet are going to be an issue over the long term.

No Defrost Cycle Needed The gas burner can also be used to remove frost or ice from the coil as needed. This eliminates the need for conventional defrosting, which is required on most heat pumps.

SECTION TWO: HEAT PUMPS

Advantages   There are several advantages to this approach:

- The back-up heat system is located outside.
- There is no internal furnace or chimney.
- There are no internal products of combustion.
- Warmed house air does not have to be used in the combustion process.
- Contaminated house air is not involved in the combustion process.
- There is very little risk of gas explosion as the burner is located outside (this is a low risk in any case).
- The system is relatively simple.
- We don't need a defrost cycle.

These systems are relatively costly to install and their long-term performance is not well-documented. The manufacturer does offer what appears to be a meaningful warranty.

These systems are not terribly common and may not be popular in your area. We encourage you to speak to local service people to determine whether systems such as this are common. An in-depth discussion of such systems is beyond our scope.

Summary

Technological changes may bring new heat pump products to market. Some heat pump systems also heat domestic water, reducing or eliminating the need for a domestic water heater.

Air Conditioning & Heat Pumps
MODULE

QUICK QUIZ 2

☑ INSTRUCTIONS

- You should finish Study Session 2 before doing this Quiz.
- Write your answers in the spaces provided.
- Check your answers against ours at the end of this Section.
- If you have trouble with the Quiz, reread the Study Session and try the Quiz again.
- If you did well, it's time for the Field Exercise.

1. Heat pumps are normally designed for the cooling load. What would allow an exception to this?

   _____

   _____

2. What are the typical cooling loads in square feet cooled per ton (in your area)?

   _____

3. What are the typical heating loads in BTUs per square foot (given in general terms in the text)?

   _____

SECTION TWO: HEAT PUMPS

4. List 12 components that are common to heat pumps and air conditioners.

   _____

   _____

   _____

   _____

   _____

   _____

   _____

   _____

   _____

   _____

   _____

   _____

5. List six components of heat pumps that are different than those that are found on central air conditioning systems.

   _____

   _____

   _____

   _____

   _____

   _____

6. An ideal heat pump arrangement typically moves more or less air across the heat transfer medium than a conventional gas furnace?

   _____

7. Is the discharge house air temperature from a heat pump higher or lower than from a gas furnace?

   _____

8. You should not operate the heat pump in the heating mode when the outdoor temperature is above _____

SECTION TWO: HEAT PUMPS

9. You should not operate the heat pump in the cooling mode when the outdoor temperature is below _____

10. Explain in two sentences why defrosting is necessary.

11. Explain in one sentence each, the two common ways of melting the frost.

12. Explain in one sentence each, the two common methods heat pumps use to detect frost on the heat pump.

13. List eight clues that would suggest that you are looking at a heat pump rather than a central air conditioning system.

SECTION TWO: HEAT PUMPS

14. List seven common heat pump problems that are in addition to the air conditioning problems discussed earlier.

_____
_____
_____
_____
_____
_____
_____

If you didn't have any trouble with this test then you are ready for the Field Exercise.

SECTION TWO: HEAT PUMPS

Key Words:

- *Compressor*
- *Evaporator coil*
- *Expansion device*
- *Condenser coil*
- *Outdoor fan*
- *Indoor fan*
- *Condensate tray*
- *Condensate lines*
- *Refrigerant lines*
- *Evaporator fan*
- *Duct system*
- *Air filter*
- *Thermostat*
- *Heat sink*
- *Heat source*
- *Reversing valve*
- *Emergency heat*
- *Accumulator*
- *Timed defrost*
- *Demand defrost*
- *Oversized*
- *Undersized*
- *Inoperative*
- *Poor location*
- *Iced-up coil*
- *Weak airflow*
- *No back-up heat*

SECTION TWO: HEAT PUMPS

▶ 5.0 INSPECTION TOOLS

Flashlight
A good light is needed to look at air conditioning and heat pump systems, particularly the coils and fans. A flashlight can also be helpful when looking at the compressor. The brighter the light, the better. Some inspectors use trouble lights (incandescent lights on cords) to inspect air conditioners and heat pumps.

Telescopic mirror
Mirrors are helpful when looking inside cabinets and plenums. Mirrors are almost always used with a flashlight.

Thermometer (with 6-inch probe-type sensor)
This can be used to check the temperature drop or rise across the indoor coil. This goes beyond the Standards but is a common test among home inspectors.

Screwdrivers and pliers
Although you shouldn't have to use tools to remove most covers, these tools are sometimes helpful to persuade stiff covers. They are also useful if you choose to go beyond the Standards and remove cabinet covers, for example.

Rags or towels
Heat pumps are often dirty. You should carry something to clean your hands when you are finished. Some inspectors carry wet wipes. Most inspectors avoid using the washroom in the house for clean up.

Tape measure
A tape measure can be useful when approximating the house square footage. However, most inspectors step off these distances.

Clamp-on Ampmeter
The ampmeter is used to measure the current draw of the compressor and compare it to the data plate. The clamp-on ampmeter can also be used to measure the current draw through electric resistance heaters to verify their operation. This is another test that goes beyond the Standards. You should only do this if you are familiar with both electricity and the use of these devices.

Your senses
As always, your eyes are your main inspection tool. You'll be using your sense of touch to feel airflow and temperature at registers and at the outdoor fan, for example. You'll also be touching Freon lines to feel temperatures there. Even without trying, your body will feel either warm or cool as you do you inspection inside the house. This will give you a general sense of the performance of the heating and cooling functions.

Tissues
Since the airflow through return registers is often hard to sense with the hand, many inspectors use tissue to check for airflow through the return grilles.

SECTION TWO: HEAT PUMPS

▶ 6.0 INSPECTION CHECKLIST

Location Legend   N = North   S = South   E = East   W = West
1 = 1st Floor   2 = 2nd Floor   3 = 3rd Floor   B = Basement   CS = Crawlspace

\multicolumn{4}{c}{HEAT PUMPS}
LOCATION	CAPACITY	LOCATION	EXPANSION DEVICE
	• Undersized for heating		• Capillary tube crimped/disconnected/leaking
	• Oversized for cooling		• Thermostatic expansion valve loose/clogged/sticking
	COMPRESSOR		**OUTDOOR FAN**
	• Excess noise/vibration		• Excess noise/vibration
	• Short cycling		• Inoperative
	• Running continuously		• Corrosion
	• Out of level		• Mechanical damage
	• Excess electric current draw		• Obstructed air flow
	• Wrong fuse or breaker size		• Dirty
	• Electric wires too small		
	• Missing electrical shutoff		**INDOOR FAN**
	• Inoperative in heating/cooling mode		• Undersized
	• Inadequate cooling		• Misadjustment of belt or pulleys
			• Excess noise/vibration
	OUTDOOR COIL		• Dirty
	• Dirty		• Dirty or missing filter
	• Damaged		• Inoperative
	• Corrosion		• Corrosion
	• Clothes dryer or water heater exhaust too close		• Damage
	• Iced up		
	• Poor location		
	INDOOR COIL		**DUCT SYSTEM**
	• Temperature split too high		• Undersized for cooling
	• Temperature split too low		• Incomplete
	• No access to coil		• Dirty
	• Dirty		• Disconnected or leaking
	• Frost		• Obstructed or collapsed
	• Top of coil dry in cooling mode		• Poor support
	• Corrosion		• Poor balancing
	• Damage		• Humidifier damper missing
			• Supply/return registers — too few
	CONDENSATE SYSTEM		• Supply/return registers — poor location
	• Pan leaking or overflowing		• Supply/return registers — obstructed
	• Dirt in pan		
	• Pan cracked		**DUCT INSULATION**
	• Inappropriate pan slope		• Missing
	• Rust or holes in pan		• Incomplete or damaged
	• Pan not well secured		
	• No auxiliary pan		**VAPOR BARRIER**
	• No float switch		• Missing or incomplete
			• Damaged

SECTION TWO: HEAT PUMPS

HEAT PUMPS			
LOCATION		**LOCATION**	
	CONDENSATE DRAIN LINE		**THERMOSTATS**
	• Leaking/damaged/split		• Inoperative
	• Disconnected/missing		• Poor location
	• Blocked or crimped		• Not level
	• No trap		• Loose
	• Improper discharge point		• Dirty
			• Damaged
	CONDENSATE PUMP		• Poor adjustment/calibration
	• Inoperative		
	• Leaking		**BACK-UP HEAT**
	• Poor wiring		• Missing
			• Inoperative
	REFRIGERANT LINES		• Poor location
	• Leaking		• Runs continuously
	• Damaged		
	• Missing insulation		
	• Lines too warm or too cold		
	• Lines touching each other		
	• Low points or improper slope in lines		

► 7.0 INSPECTION PROCEDURE

The inspection procedure is very similar to that used for central air conditioning. It's outlined below.

A. VISUAL INSPECTION

Outdoors

1. Find the outdoor unit and check the cabinet for rust or damage.
2. Make sure that the outdoor unit is level within 10 degrees.
3. Make sure the unit is not partially buried in the soil.
4. Visually inspect the outdoor coil for dirt, lint or other obstructions and for corrosion or fin damage.
5. Look for a clothes dryer, water heater or high efficiency furnace vent within 6 feet of the unit.
6. What condition is the outdoor fan in?
7. Is the airflow obstructed? (You should have 1 to 3 feet on the intake side and 4 to 6 feet on the discharge.)
8. Is there evidence or corrosion or damage?

Note: The compressor may be outdoors or in a separate cabinet indoors.

9. Record the unit size, age, compressor Rated Load Amperage (RLA), maximum fuse size and minimum circuit ampacity from the data plate.
10. Check the unit size against the square footage of the house (general rule).
11. Look for an outdoor electrical disconnect within sight of and readily accessible to the outdoor units.
12. Check the size of the fuses or the breakers.
13. Check the circuit ampacity against the wire size. Check the condition of wires and connections. Check that outdoor wires are suitable for outdoor use.
14. Make sure that the outdoor coil is –

    a. well off the ground so it won't be buried in snow (in snowy climates), and so condensate can drain away freely

    b. in an area where air will not be recirculated

    c. not exposed to the prevailing wind, or

    d. not below the drip edge of the roof.

15. For those who go beyond the Standards –

    a. Shut off the power

    b. Open the cabinet

    c. Record the age of the compressor

    d. Check the reversing valve for obvious damage, leakage or corrosion

    e. Check the accumulator for obvious damage, leakage or corrosion

    f. Look for a bulging capacitor

    g. Look for oil in the cabinet indicating refrigerant leakage

    h. Look for dirt, fin damage and corrosion

    i. Check the electrical connections

    j. Close the cabinet unless you are going to check the electrical current draw when the system is operating

    k. Turn the power back on

Indoors and out

1. Check the refrigerant lines for insulation, oil deposits (indicating leakage) and evidence of crimping or mechanical damage. Refrigerant lines should not touch each other.

Indoors

1. Check for service access to the plenum coil. If it's available, open the access.

2. Check the size of the coil, if visible, and compare it to the outdoor coil size.

3. Check the indoor coil for dirt, damage, corrosion or frost. Pay close attention to the upstream side. If the unit has been operating in the cooling mode, make sure that the entire coil is wet.

4. Check the expansion device (capillary tube or thermostatic expansion valve) for evidence of leakage or mechanical damage. There may be two expansion devices, one near each coil.

5. Check the condensate (and auxiliary) drain pan for evidence of leakage, dirt, rust, improper slope or poor securement. If there is no auxiliary pan, and the coil is above finished living space, a high water-level cut-out (float switch) should be present.

6. Check the primary and secondary (if applicable) condensate line for a trap, evidence of leakage, blockage, and appropriate discharge point. If there is a pump, check the wiring to the pump. Is pump grounded? Is there any evidence of leakage around the pump?

    Note: Turn the power off and remove the blower cover.

7. Check the blower size against the furnace data plate (if applicable). Check the fan blades for dirt.

8. Check for a loose or worn belt or misalignment of the belt and pulleys.

9. Check for the presence and condition of an air filter.

SECTION TWO: HEAT PUMPS

10. If the auxiliary heat is electric, make sure the heat is after (downstream of) the indoor coil. If the back-up heat is a gas or oil furnace, the back-up heat exchanger should be upstream of (before) the heat pump.

Note: Replace the blower cover and turn the power back on.

B. OPERATING TESTS

Testing In Heating And Cooling Modes

While some inspectors do switch modes during certain weather conditions, we do not. We test the heat pump in the cooling mode if that's where the thermostat is set, and the outdoor air temperature is above 65°F. We test the heat pump in the heating mode if that is where it's set, as long as the outdoor temperature is below 65°F.

Running the heat pump in cooling mode when the temperature is below 60°F to 65°F risks damaging the system, as does running the unit in the heating mode above 65°F.

In cooling mode

Testing a heat pump in the cooling mode is exactly the same as testing an air conditioner.

Note: If the furnace has been running, wait at least ten minutes after the house air fan stops before testing the air conditioning. This allows the refrigerant pressure to equalize through the system and avoids damaging the compressor.

1. Determine if the system has had power for at least 12 hours.
2. If the outdoor temperature is above 65°F (and has been for at least 12 hours or so), turn the thermostat down to its lowest setting. Listen for the house air fan coming on.
3. Check the outdoor fan for unusual noise and vibration.
4. Check the compressor for unusual noise and vibration. Watch for short cycling.
5. After the system has been running for 15 minutes, check the air coming off the outdoor coil with your hand. It should be warmer than the ambient air.
6. Check the temperature of the refrigerant lines.
7. If you go beyond the Standards and use an ampmeter, measure the electric current flow through the compressor. Compare it to the compressor Rated Load Amperage (RLA). It should be roughly 60 to 90 percent of the RLA. Replace cabinet cover.
8. Check the temperature difference (split) across the indoor coil.
9. Check the indoor coil for frost.
10. Check the condensate line to ensure water is flowing out. Look for leaks in the condensate pan and line.

SECTION TWO: HEAT PUMPS

11. If there is a float switch for the condensate pan, hold the float up so the unit thinks that the pan is full. The unit should shut off. This may take about five minutes. Release the float. The unit should re-start. This may take five to seven minutes.
12. If there is a condensate pump, check that it is working properly.
13. Replace or close the access cover for the coil.
14. Check the house air fan for unusual noise and vibration.
15. Check the ducts for leakage, missing insulation and compromised vapor barriers. Are ducts in attics and crawlspaces sweating?
16. Check the airflow at supply and return registers throughout the house.
17. Return the thermostat to its original position.

2. In Heating Mode

Testing a heat pump in the heating mode is done as follows:

1. Check that the thermostat is set to the heat mode.
2. Check that the outdoor temperature is below 65°F.
3. Complete the visual checks as outlined earlier.
4. Raise the thermostat until the compressor starts. The outdoor fan and indoor fan should also start. If it's a typical two-stage thermostat, just raise the setting enough to turn the heat pump on. If you raise it too much, the back-up heat may come on.
5. Allow the unit to run for 10 to 15 minutes.
6. Listen to the compressor and the fans for unusual noise or vibration.
7. Check the temperature rise if you are going to go beyond the Standards. You should be looking for a temperature rise across the indoor coil of about 20° to 30°F depending on the outdoor temperature. This means if the return air coming back from the living space is 70°F, the air leaving the supply plenum may only be 90°F.

   Note: this assumes that the back-up heat is not operating.

When the heat pump has been operating for a few minutes, the air coming off the outdoor fan should be colder than the outdoor air.

8. Check the temperature of the Freon lines. You need to know the location of the compressor and the expansion device. Let's assume R-22 is the refrigerant and the outdoor temperature is 35°F.
   a. The liquid line upstream of the expansion device should be warm (about 105°F to 115°F)
   b. The liquid line downstream of the expansion device should be cold (15°F to 20°F)
   c. The suction line upstream of the compressor should be cold (25°F to 30°F)
   d. The suction line downstream of the compressor should be hot (160°F to 170°F)

SECTION TWO: HEAT PUMPS

9. Check that the house temperature is rising. Discharge temperatures at supply registers should be roughly 85°F to 100°F. This will vary with the outdoor temperature.

10. If the heat pump does not come on,

    a. there may be an operational problem

    b. the heating mode of the heat pump may have been abandoned

    c. the heat pump may be locked out because the outdoor temperature is too low, or

    d. the thermostat may be set to **Emergency Heat** (You can check this.

11. Check to see whether the back-up heat is on. In testing the back-up heating system, you should use the inspection techniques described in the Heating Modules for the appropriate system.

12. Electric resistance heat can work simultaneously with the heat pump. If there is more than one electric resistance heater, the heaters may come on at different times because of sequencers that energize individual elements about 30 seconds apart. The unit may have staging, which only introduces enough electric heat to keep the house temperature from falling. Sometimes the thermostat has to be raised all the way to make all of the elements come on.

13. If the back-up heat does not come on,

    a. it may be inoperative, or

    b. the outdoor temperature may be too mild

14. On most heat pumps, the thermostat can be set to the **Emergency Heat** mode. This should bypass the heat pump, shutting it down, and should activate the back-up heat. The back-up heat source can be tested using the **Emergency Heat** setting.

15. Return the thermostat to the **Heating** mode. If the heat pump had been running, it should start again, and the back-up heat should shut down.

16. Check the airflow through supply and return registers.

17. Turn the thermostat back to its initial position. The heat pump (and back-up heat) should shut off.

18. If there is an ice build-up on the outdoor coil, the heat pump may not work in the heating mode immediately. It may be in defrost cycle. If the frost build-up is more than $\frac{1}{8}$ inch, the defrost cycle may not be working. If a heavy frost accumulation is noted on the outdoor coil, it should be checked several times during the inspection to ensure that the frost is removed during regular defrosting.

SECTION TWO: HEAT PUMPS

Air Conditioning & Heat Pumps
MODULE

FIELD EXERCISE 1

☑ INSTRUCTIONS

You should complete everything up to and including the Inspection Procedure before starting this Exercise. For this Field Exercise, you will look at some houses with heat pumps. The more you look at the better, but look at between five and ten, if you can. Ideally, you are looking for heat pumps of different ages and different manufacturers.

The goal here is to help you identify, inspect and test heat pumps in the field and to get some practice looking for the common problems.

You should allow yourself about 45 minutes for each heat pump you look at during this Field Exercise.

We are going to break the exercise down into three parts:

- Exercise A — A visual inspection
- Exercise B — An operating inspection in either the heating or the cooling mode
- Exercise C — Research into the kinds of systems and the kinds of problems common in your area

We will focus on air-to-air heat pumps. If other types of heat pumps are common in your area (as you will discover when talking to the experts), you will need to develop inspection procedures for those as well. In most cases, they will be similar to the inspection and testing of an air-to-air system.

SECTION TWO: HEAT PUMPS

Exercise A – Visual inspection

1. Let's start by making sure that it is a heat pump.
   a. Can you find the data plate? It is usually on the outdoor coil cabinet.
   b. Is it clear from the data plate that it is a heat pump? If not, take note of the manufacturer's name and model number so you can contact the manufacturer.
2. Does the thermostat have an Emergency Heat setting?
   a. Is the thermostat a two-stage unit? On mercury bulb-type thermostats, that means that there will be two bulbs.
3. Are both of the Freon lines insulated?
4. Can you find a reversing valve in the cabinet?
5. Are there two expansion devices (one close to the indoor coil and one close to the outdoor coil)?
6. Is the compressor indoors?
7. Is there an outdoor temperature sensor (thermostat) connected to the control wiring?
8. Is it below 65°F and is the unit operating in the heating mode?
9. Is there frost on the outdoor coil?

Now, let's look at what might be wrong. The Inspection Checklist and Inspection Procedure may be helpful here.

10. Look for all of the problems that we talked about with respect to central air conditioning systems. Also, look for the following:
    a. Is the system oversized for cooling? It may well be, if too much attention was paid to making the system large enough to heat the home.
    b. Is the outdoor coil –
       (1) under the drip line of a roof?
       (2) in a spot where snow drifts may accumulate?
       (3) in a spot where condensate cannot run freely away?
       (4) in an enclosed area where air cannot move freely?
       (5) exposed to the prevailing wind?
    c. Is the outdoor coil iced up?
    d. Is there back-up heat? If the back-up heat is electric, it should be after (downstream of) the heat pump. If the back-up heat is a gas or oil furnace, the back-up heat exchanger should be upstream of (before) the heat pump.

Exercise B – Testing the system operation.

1. Follow the Inspection Procedure in Section 7.0 for testing air-to-air heat pumps.
2. We recommend only testing in the heating or cooling mode, depending on the season. We do not recommend changing the operating mode unless the inspector has expertise in this area (this goes beyond the Standards).

During the operating test, there are many possible scenarios that you could come across. Life is simpler if we do not have back-up heat, but most systems do. The operating possibilities are different depending on whether it is electric or gas or oil, for example. Let's look at electric first and assume that the system is in the heating mode (outdoor temperature is below 65°F).

3. What are the possibilities?
   a. The heat pump may be operating
   b. The heat pump and the back-up heat may be operating
   c. only the back-up heat may be operating

When the temperature is above the balance point, you might expect to find only the heat pump running. When the temperature is below the balance point, but the COP is greater than 1.0, you might expect to find the heat pump and the back-up heat operating. If the temperature is below the balance point and below the point at which the COP is 1.0, you might find only the electric back-up heat operating.

The back-up heat should not operate when the heat pump is operating in mild weather. Improper wiring can cause electric heaters to operate whenever the system is working. This is inefficient and defeats the purpose of the heat pump.

If the thermostat is a two-stage thermostat and you turned it up dramatically from its original setting, you may have caused the back-up heat to come on even though it would not normally operate at that temperature.

If only the back-up heat is operating, it may also be because the heat pump itself is inoperative or because the thermostat is set to Emergency Heat.

This discussion should make you cautious about criticizing the operation of the heat pump and the back-up heat system without knowing what the balance point of this system is and at what point the COP is equal to 1.0. It may be difficult to determine if the system is operating properly or not. It is probably best to report your observations and, if the operating mode seems unusual for the outdoor temperature, simply raise a question.

4. Let's look at the possible scenarios for add-on heat pumps (gas or oil back-up systems).
   a. Only the heat pump is on
   b. Only the back-up is on
   c. The heat pump and the back-up heat are on

SECTION TWO: HEAT PUMPS

Add-On Heat Pump Controls The add-on heat pump is more complicated because the heat pump and the back-up heat cannot be on at the same time. Three different approaches are used.

a. The simplest control system is to shut off the heat pump when the balance point is reached. If the heat pump can't provide all of the heat that is needed, the furnace takes over. An outdoor thermostat senses the temperature at which the balance point occurs and shuts off the heat pump at this point, allowing the furnace to take over. Above the balance point setting, the furnace won't come on unless the system is switched to the Emergency Heat position.

b. The **Philadelphia method**, or **unrestricted method** as it is sometimes called, keeps the heat pump running even when the system is below the balance point. With this system, the heat pump comes on every time that there is a call for heat and only shuts off when the temperature in the home continues to fall. At this point, the furnace takes over. The next time the thermostat calls for heat, the heat pump comes on. If the temperature continues to fall, the furnace comes on and the heat pump shuts off.

c. The **plenum-controlled system** has a two-stage thermostat with the heat pump controlled on the first stage and the furnace on the second stage. Above the balance point, the heat pump operates controlled by the first stage of the thermostat. Below the balance point, the second stage of the thermostat turns on the furnace. The heat pump may or may not come on briefly while the furnace is warming up. If the heat pump is on while the furnace is warming up, it will shut off when the temperature in the plenum hits about 95°F.

The furnace will run until the second stage of the thermostat is satisfied at which point it turns off. As the air temperature in the plenum drops, the heat pump will come back on at about 90°F. The heat pump delivers some heat into this relatively cool air. At very low outdoor temperatures, the heat pump is sometimes locked out with an outdoor thermostat.

Again, consider what should be operating. If the temperature is above the balance point (32°F is a common balance point, but this number can vary dramatically), the heat pump alone would typically be running. If you are below the balance point, the furnace alone should usually be running.

If the temperature is relatively mild and the heat pump is not running but the furnace is:

- the heat pump may be defective
- it may have been abandoned in the heating mode
- it may have a lock-out temperature that is set unusually high, or
- the thermostat may be set to Emergency Heat

Again, be wary of criticizing the system unless you have some expertise in this area. If the system seems to be performing outside of normal parameters, call for further investigation. Remember that you are not there to trouble shoot systems.

SECTION TWO: HEAT PUMPS

5. If the system is operating in the cooling mode, we'll treat it as an air conditioner. We discussed the inspection and testing procedures in the first section of this Module.

Exercise C – Speaking to the experts

Talk to local heating contractors about common heat pump systems in your area.

1. Are air-to-air heat pump systems common?
2. Are they typically successful?
3. Is their installation or operation subsidized by utilities or other agencies? (This has been common in some areas.)
4. Are air-to-air heat pumps the most common?
5. Are others found, such as ground-source or water-source heat pumps?
6. What are the most common operating problems found with heat pumps?
7. Are undersized duct systems an issue? What about relatively low temperatures at the supply registers?
8. Are there reliability problems?
9. Are operating costs significantly lower than other heating systems?
10. Are there hybrid systems that are used to heat domestic water, for example?
11. What are the most common back-up heat arrangements?
12. What are common poor locations for outdoor coils?
13. Are there any issues that you should be aware of while inspecting and testing heat pumps?
14. Is it common for heat pumps to have the heating mode abandoned?
15. What liquids do water source heat pumps use in closed systems?
16. What are common life expectancies for heat pumps?

When you have completed the Field Exercise, you are ready for the Final Test.

SECTION TWO: HEAT PUMPS

► 8.0 BIBLIOGRAPHY

TITLE	AUTHOR	PUBLISHER
Home Heating & Air Conditioning Systems	James L. Kittle	McGraw Hill
Heating, Ventilating and Air Conditioning Library, Volumes I, II & III	James E. Brumbaugh	Macmillan
Various papers	Compilation	Heating, Refrigerating and Air Conditioning Institute of Canada
1997 ASHRAE™ Handbook Fundamentals	Compilation	American Society of Heating, Refrigerating and Air-Conditioning Engineers, Inc.
1996 HVAC Systems and Equipment	Compilation	American Society of Heating, Refrigerating and Air-Conditioning Engineers, Inc.
1995 HVAC Applications	Compilation	American Society of Heating, Refrigerating and Air-Conditioning Engineers, Inc.
Advanced Air Conditioning — Beyond the Basics (a paper)	Mark Cramer	Mark Cramer
Heat Pump Manual	Compilation	The Electric Power Research Institute
Heat Pump Systems	H.J. Souer and R.J. Howell	John Wiley and Sons
Heat Pumps: An Efficient Heating and Cooling Alternative	D. McGuigan	Garden Way Publishing Company
Heat Pump Primer	Compilation	Ontario Hydro

SECTION TWO: HEAT PUMPS

► ANSWERS TO QUICK QUIZZES

Answers to Quick Quiz 1

1. Condenser

2. Evaporator

3. c.

4. Near the compressor in a cabinet, either indoors or outdoors

5. c.

6. b.

7. It would cool the house down too fast without dehumidifying the air, resulting in a cool, damp environment.

8. COP is the ratio of the amount heat being delivered to the indoors divided by the amount of electricity being consumed by the heat pump.

9. 3.0

10. When COP drops below 1.0

11. The balance point is the temperature at which the rate of heat delivered by the heat pump is equal to the heat loss to the outdoors.

12. False

13. True

14. a) The heat pump gets the first chance to warm the air, allowing the furnace to rest until it's needed.
    b) In the summer, when the heat pump cools the air before it gets to the furnace, it would condense moisture out of the air on the burner side of the heat exchanger, causing the heat exchanger to corrode.

15. In winter, the air moving across the coil is cooled and may become saturated. If the temperature is below freezing, frost instead of condensation develops on the coil.

16. 1. Water source
    2. Ground source

SECTION TWO: HEAT PUMPS

Answers for Quick Quiz 2

1. We could use a two-stage compressor, which could have a lower output in the summer than in the winter.

2. 1. 450 to 700 square feet – southern U.S.A.
   2. 700 to 1000 square feet – moderate climates

3. 1. 40 to 60 BTUs per square foot in older homes
   2. 25 to 40 BTUs per square foot in newer homes

4. 1. Compressor
   2. Indoor coil
   3. Expansion device
   4. Outdoor coil
   5. Outdoor fan
   6. Condensate tray and lines
   7. Refrigerant lines
   8. Indoor fan
   9. Duct system
   10. Duct insulation
   11. Air filter
   12. Thermostat

5. 1. Reversing valve
   2. Both refrigerant lines may be insulated
   3. May have an outdoor air temperature sensor
   4. Emergency heat setting on the thermostat
   5. Two-stage thermostat
   6. Compressor may be indoors
   7. The heat pump may be a triple split system

6. More air is moved for a heat pump than a furnace

7. Lower for a heat pump than a furnace

8. 65°F

9. 65°F

10. When the heat pump is operating in cold weather, the temperature of the coil is cooler than the air around it. The air becomes saturated, and frost develops on the coils, reducing the efficiency of the unit.

11. 1. Electric heaters can melt the ice.
    2. The heat pump reverses, operating in cooling mode with the outdoor fan off. This heats the outdoor coil, melting the ice.